T0141154

Millennial Landscape Change in Jordan

Millennial Landscape Change in Jordan

Geoarchaeology and Cultural Ecology

Carlos E. Cordova

The University of Arizona Press   Tucson

The University of Arizona Press
© 2007 The Arizona Board of Regents
All rights reserved

Library of Congress Cataloging-in-Publication Data
Cordova, Carlos E., 1965–
Millennial landscape change in Jordan : geoarchaeology
and cultural ecology / Carlos E. Cordova.
p. cm.
Includes bibliographical references and index.
ISBN-13: 978-0-8165-2554-6 (hardcover : alk. paper)
ISBN-10: 0-8165-2554-4 (hardcover : alk. paper)
1. Jordan—Geography. 2. Landscape changes—Jordan.
3. Archaeological geology—Jordan. I. Title.
G76.5.J6C67 2007
304.2095695—dc22
2006028194

∞
Manufactured in the United States of America on acid-free,
archival-quality paper containing a minimum of 50% post-
consumer waste and processed chlorine free.

12   11   10   09   08   07   6   5   4   3   2   1

Contents

Figures

Tables

Preface

The territory of the Hashemite Kingdom of Jordan occupies four of the five vegetation provinces of the Middle East (fig. P.1). The Mediterranean province in Jordan is a narrow belt cut off from the rest of the Mediterranean realm by the Jordan Rift Valley. This narrow belt is surrounded by drylands, a situation that results in a unique mosaic of cultural and natural landscapes. The Irano-Turanian steppe is a transitional landscape linking the Mediterranean region with the desert. The term *steppe* here refers to a climatically induced treeless landscape, not to the typical midlatitude steppe of Eurasia. Although the Jordanian steppe seems to have once been more vegetated and diverse, it is currently formed mainly by scrub.

The Saharo-Arabian region is the true desert, although it is not as monotonous as most people believe. The Jordanian deserts present different lithologies (rock, sand, and salt pans), numerous oases, and different variations. The two main desert areas correspond to the Saharo-Arabian and Sudanian provinces, which present an interesting combination of Eurasian and African biota. Thus, this book focuses on the natural and cultural importance of these four landscape zones and how geoarchaeology and cultural ecology help us understand them.

The subjects of my research in Jordan were inspired by dozens of readings on the geography and archaeology of the Middle East. But my interest in the topic of this book was mainly stirred by ideas set down by travelers and archaeologists who visited the region in the late eighteenth century and early nineteenth century. George Perkins Marsh (a geographer, diplomat, and traveler who visited the Middle East in the late 1800s) proposed that the unfruitful appearance of the rural landscapes of this region was a result of neglect by the populations who had inhabited the region for millennia (see Lowenthal 2000). Ellsworth Huntington (1911) also contributed some ideas to explaining the nature of the Transjordanian landscape, although he pointed to climate change as the main agent transforming the landscape into a wasteland. Walter Clay

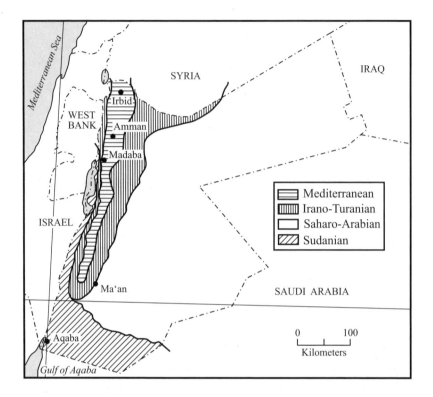

Figure P.1 Distribution of floristic provinces in Jordan (based on data from Zohary 1973 and Al-Eisawi 1985).

Lowdermilk (1944) and Adolf Reifenberg (1955) blamed the nomadic populations for the devastation of vegetation and soil. I realized that all the writings by these scholars lacked a solid argument based on empirical data. Their arguments were based not on paleoecological data (faunal and floral remains, soils, geology, etc.) but on the appearance of a land ravaged by socioeconomic and environmental mismanagement, nomadic raids, and droughts during the last decades of Ottoman rule.

Despite the pathetic appearance of the Transjordanian landscapes during the first third of the twentieth century, the archaeologist Nelson Glueck viewed it in a different way. During his archaeological survey of Transjordan (Glueck 1939), he reported remains of aqueducts, canals, dams, terraces, and other material as evidence of past economic prosperity. The tide then began to change, as most archaeologists and geographers working on the comparatively recent past became interested in exploring evidence for periods of social and economic boom and their termination. Thus, this research focus prompted more interest in the reconstruction of the relationship between societies and their environments in antiquity.

In addition to information recovered through various researchers working on the paleoenvironment and human populations in prehistoric and historic times, I am presenting results recovered through field research during the period 1995–2005 in various regions of Jordan. The purpose of this book is twofold: it is an introduction to the physical geography of Jordan for archaeologists, historians, geographers, and other scholars of social science and humanity; it also delineates some research possibilities in the areas of geoarchaeology and cultural geography in Jordan. This book is particularly directed toward students participating in archaeology field schools in Jordan and geography students interested in environmental issues in the Middle East.

Organization of the Book

Chapters 1 and 2 introduce the reader to the scope of the research and the physical scene. Chapters 3 and 4 describe the main subjects: woodlands, steppes, and deserts. Chapter 5 is a review of the current geoarchaeological and paleoecological records and an assessment of sources of paleoenvironmental information that present potential for landscape reconstruction. Chapter 6 is a review and discussion of millennial landscape change in Jordan, focusing on climate change, cultural development, and human-induced landscape transformation. Although this chapter focuses on the period encompassing the Last Glacial Maximum (roughly 20,000–18,000 years BP) to the end of the first phase of urbanization (ca. 4000 BP), references are also made to previous and later periods. Chapter 7 presents the discussion of the most significant cultural ecological issues in paleoenvironmental research in Jordan. At the end of the book is a glossary of technical terms, which contains definitions for technical words in geoarchaeology, ecology, and Quaternary science.

Editorial Notes

Reported Ages and Climatic and Cultural Periods

Uncalibrated dates are reported as radiocarbon years before present, or BP (e.g., 2890 ± 50 ^{14}C years BP). Calibrated radiocarbon dates are reported as cal. years BP (e.g., cal. 2635 BP) or cal. years BC or AD (e.g., cal.

5552 BC; cal. 1010 AD). In some cases, calibrated dates are reported as a range (e.g., cal. 1998–2017 BC), which is the two-sigma standard deviation based on 95 percent probability that the true age falls within that range. When dates are rough estimations, they are reported as ka BP (*kilo annum* before present, where 1 ka = 1,000 years; e.g., 116 ± 5.3 ka BP) or Ma BP (where 1 Ma = 1,000,000 years).

Climatic periods of the Pleistocene are often based on conventional international standards such as Marine Isotope Stages (MIS), Last Glacial Maximum, Younger Dryas, and Early Holocene Climatic Optimum, in which case a bibliographic source is provided. Definitions of such terms are also included in the glossary. Other time subdivisions referred to throughout the book are the Terminal Pleistocene (which spans 20,000–10,000 BP), including the Younger Dryas (10,800–10,000 BP), and the Holocene, encompassing the Early Holocene (10,000–6000 BP), Middle Holocene (6000–4000 BP), and Late Holocene (4000 BP to the present).

The cultural chronological periods are based on the general Levantine nomenclature (or their Transjordanian equivalents when appropriate). For prehistoric periods, the chronological nomenclature is based on Henry 1997b, Olszweski 2001, and Rollefson 2001b. Subdivisions into lithic industries and assemblages are often referred to a bibliographic source. For historic periods, the divisions and ages draw on the contributions in MacDonald, Adams, and Bienkowski 2001. The summary in figure P.2 is intended to be a general guide. More detailed chronologies for the Epipaleolithic, Neolithic, Chalcolithic, and Early Bronze Age are contained in figures and tables in chapter 6.

Place-Names and Political Boundaries

The designation "Jordan" refers to the present territory of the Hashemite Kingdom of Jordan. "Transjordan" often refers to the land east of the Jordan River, but in this book the term is used to designate the territory of Jordan when referring to periods preceding the creation of the Hashemite Kingdom in 1946. The lands west of the Jordan River are referred to here as "Palestine," regardless of the modern political and national connotation of this name. In some instances, Palestine is referred to as "Israel and the Palestinian territories," in which the latter name refers to the West Bank and the Gaza Strip. Boundaries traced on the maps of this book are not intended to favor a particular ethnic or national group. They are used here only as a geographic reference. After all,

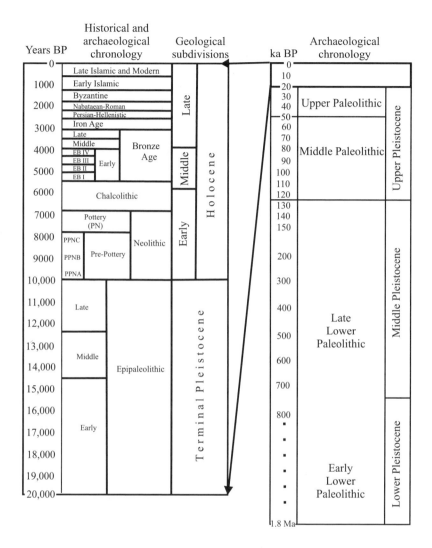

Figure P.2 Chronological cultural sequence for Jordan. The abbreviations "ka" and "Ma" indicate thousands of years and millions of years, respectively.

the facts discussed in this book refer to periods preceding the formation of modern political boundaries and states. The designation "Levant" is used here to refer to the region encompassed by the modern territories of Syria, Lebanon, the State of Israel, the Palestinian territories, Jordan, and the Sinai Peninsula. "Southern Levant" excludes Syria and Lebanon.

The transliteration systems of Arabic place-names into English-language characters vary considerably in the archaeological, geographical, and historical literature. Therefore, most place-names here do not conform to a specific system of transliteration; they instead reflect the most common spelling, as used on maps and road signs. Table P.1 illustrates the pronunciation of Arabic words used in this book.

In chapter 6, the designations "Northern Moab" and "Southern Moab" (or Moab proper) are used. The geographic separation between the two regions is marked by Wadi al-Mujib (the biblical Arnon River). Moab is the geographical designation for a biblical-historical region that is often used in the historical and archaeological literature. The name corresponds to the designation "'Arḍ al-Mu'abīyīn" in Arabic. In addition, the designations "Gilead," "Ammon," and "Edom," like "Moab," are used throughout the book, mainly when citing references from historical and archaeological sources. Biblical names of regions and main geographic structures are occasionally mentioned, when appropriate. However, for equivalents between modern and biblical place-names, I encourage the reader to consult MacDonald 2000. The more complicated modern traditional names (cf. Salibi 1998) for the regions of Jordan in Arabic are not used in this book, because they are not popular in the Western literature.

For the purpose of geographic differentiation of areas with particular characteristics, this book presents some geographic designations never used before, such as "Transjordanian Mediterranean Belt" (TJMB), which is employed here to refer to the region of Jordan with Mediterranean-type climate, vegetation, and agriculture. This region roughly corresponds to the Mediterranean vegetation territory in figure P.1.

Geological, geomorphological, and biogeographical designations conform to those used in Zohary 1973, Bender 1974, Al-Eisawi 1985 and 1996, and Macumber 2001. However, nonconventional geographic designations used in this book include, for instance, the "Levantine Loess Belt" and "Khan az-Zabib Loess," as well as "Mafraq Plateau," among other designations that are used to subdivide formerly named physiographic units.

The designation "the Plateaus" includes the Irbid Plateau, the Northern Moab Plateau (subdivided into the Madaba and Dhiban plateaus), the Southern Moab Plateau (or Karak Plateau), and the Tafila–Ras an-Naqb Plateau. "The Escarpment" refers to the western limit of these plateaus and the western flank of the Ajlun Mountains. "The Western

Table P.1   Pronunciation Guide for Arabic Words

Consonant sounds

th, as *th* in "thin"

dh, as *th* in "then"

kh, as *ch* in "loch"

r, rolled *r* as in Spanish

gh, as the French *r*

s, as in English

ʿ (ʿain). No equivalent in English. This is a guttural sound produced by compression of the throat and expulsion of air.

ʾ Glottal stop.

ḍ Emphatic *d*-sound formed by placing the tongue against the lower palate instead of against the teeth.

ẓ Emphatic *dh*-sound formed by placing the tongue against the lower palate instead of against the teeth.

ṭ Emphatic *z*-sound formed by placing the tongue against the lower palate instead of against the teeth.

ḥ Emphatic *h*-sound pronounced with a strong expulsion of air from the chest.

ṣ Emphatic *s*-sound formed by placing the tongue against the lower palate instead of against the teeth.

Vowel sounds

ā, as *a* in "far"

ī, as *ee* in "fee"

ū, as *oo* in "noon"

a, as *a* in "lamb"

i, as *i* in "pit"

u, as *oo* in "crook"

Highlands" include all the plateaus and the Ajlun Mountains. "Al-Ghor" refers to the Jordan Valley and the Dead Sea basin. "The Rift Valley" includes the Jordan River valley, the Dead Sea basin, and the Wadi Araba.

Acknowledgments

Although the ideas that led to my writing this book originated in different academic contexts, they all converge in this volume. One of the fundamental ideas underlying my approach came from my interaction with Karl Butzer, my doctoral supervisor. Although Karl never encouraged me to write a book on Jordan, he did put me on the path to synthesizing information derived from multiple sources. His writings on cultural ecology and geoarchaeology from Egypt, Sub-Saharan Africa, the Mediterranean, and Mexico are wonderful examples of data synthesis and interpretation.

In addition, while carrying out geoarchaeological research, I realized the common ground between the fields of archaeology-anthropology and geography-geosciences, which requires cooperation and a dialogue. For this reason, I am deeply indebted to all the archaeologists with whom I have interacted during the past twelve years of research in the Middle East and other parts of the world. Learning the local names of plants and talking to Bedouins and farmers in their language gave me a strong insight into what plants mean to the modern people and what they meant to past civilizations. I greatly appreciate the contributions of all those who spoke with me and gave me links to connect culture with environment.

For their interest and their encouragement to write this book, I would like to thank a number of archaeologists working in Jordan, including Don Henry, Gary Rollefson, Phil Wilke, Leslie Quintero, April Nowell, Michael Bisson, Chris Foley, Suzanne Richard, and Jesse Long. I would also like to thank archaeologists Michele Daviau, Lisa Maher, Maysoon al-Nahar, Dawn Cropper, Tim Harrison, Bruce Routledge, Benjamin Porter, and Steve Savage for their cooperation and interest in my research. I have also greatly benefited from discussions with archaeologists Nancy Coinman, Geoff Clark, Arlene Rosen, Steve Rosen, and Burton MacDonald.

Those colleagues working in geology, geoarchaeology, and paleoecology who positively influenced this project were Caroline Davis, Tina Niemi, Yahya Farhan, Nizar Abu-Jaber, and Muheeb Awawdeh. I would also like to thank Moshe Inbar, Yehuda Enzel, Tony Wilkinson, Charles Frederick, Karl Butzer, Tim Beach, Vaughn Bryant, and Dafna Kadosh, all of whom work in the Levant and the eastern Mediterranean region. My special thanks to Henk Woldring and the late Sytze Bottema for allowing me to work with their pollen collection housed at the University of Groningen.

Local support for this project was kindly provided by the American Center of Oriental Research (ACOR) through its facilities in Amman and Madaba. Therefore, I would like to thank Pierre and Patricia Bikai, the former director and associate director of the center, as well as Kurt Zamora, Carmen (Humi) Ayoubi, Kathy Nimri, and Nisreen Shaikh, and the rest of the wonderful ACOR staff.

I would also like to express thanks to Fawwaz al-Kharaysheh, director of the Department of Antiquities of Jordan, and Mohammad Najjar for support in the projects I worked for. Also, of great help in the Department of Antiquities were Huda Kilani and many other representatives working for projects. Other scholars in Jordan that I would like to thank are Dawud al-Eisawi of the University of Jordan and Rami Khouri, who works as a journalist and advisor to the Ministry of Tourism and Antiquities of Jordan.

Research that led to the compilation of data for this book was carried out with funding from the Committee for Research and Exploration of the National Science Foundation (Grant Number 6540-99), start-up grants from the Association of American Geographers and the College of Arts and Sciences of the Oklahoma State University, and the numerous grants received by my collaborators in Canada, the United States, and Jordan.

Additionally, I would like to thank Michael Larson, head of the Cartography Lab at the Geography Department of Oklahoma State University, and assistants Lori Flynn, April Gillilan, and Jeremy Odenwald for the drafting of most of the figures in this book. Parts of the manuscript were read and commented on by Caroline Davis, Dale Lightfoot, Gary Rollefson, Don Henry, Adrienne Sadovsky, Paul Lehman, and Jess Porter.

Finally, I would like to thank the editorial staff of the University of Arizona Press—particularly Allyson Carter, Christine Szuter, and Al Schroder—and freelance editor Sally Bennett for their participation in the editing and publication of this manuscript.

Millennial Landscape Change in Jordan

1

Approaches to the Study of Ancient Jordanian Landscapes in the Near Eastern Context

On the Fertile Crescent's Edge

In *The Other Side of the Jordan*, Nelson Glueck (1970) indicated that the largest archaeological remains of continually occupied cities are located in the region of Transjordan with the most benign climate. He referred to the narrow strip of land with Mediterranean-type climate occupying the highlands east of the Jordan River. However, continually occupied cities also exist in the lowlands of the Jordan Valley and the Dead Sea Plain, which, although arid, benefit from the waters of the Jordan River and the streams descending from the highlands. For this reason, the Transjordanian highlands and the Jordan Valley are part of the western end of the Fertile Crescent, the cradle of ancient Near Eastern civilization (fig. 1.1).

The Nabataeans, the major civilization that flourished within the present territory of Jordan, extended along the very edge of the Fertile Crescent. Their cities occupied the steppes and deserts adjacent to the Mediterranean region of the southern Levant, the Sinai, and the northwestern part of the Arabian Peninsula. Unlike other Near Eastern civilizations, the Nabataeans flourished in a dry land with no major river and no sea. Nevertheless, they built cities, roads, canals, and dams and developed a complex society.

The Nabataeans were not the only ones who built cities in Transjordan. Long before them, the peoples of the Early Bronze Age societies built cities and used irrigation. The societies of the Hellenistic, Roman, Byzantine, Early Islamic, and Mameluke periods also built numerous urban centers, cobbled roads, canals, qanats, aqueducts, dams, and terraces, which in subsequent periods fell into abandonment and oblivion until travelers and archaeologists such as John Lewis Burckhardt (1822), Ulrich Jasper Seetzen (1854), Henry Baker Tristram (1873), Selah Merrill (1883), and Charles Montagu Doughty (1936) rediscovered them and reported them to the Western world. In the barren and impoverished Transjordan of the late nineteenth and early twentieth century, these

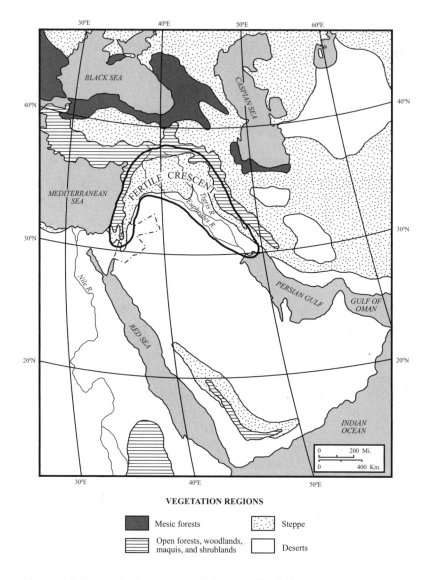

VEGETATION REGIONS

■ Mesic forests	⋮ Steppe
☰ Open forests, woodlands, maquis, and shrublands	□ Deserts

Figure 1.1 Jordan in the context of the main landscape regions (or floristic provinces) and the Fertile Crescent.

travelers and archaeologists were amazed by these dusty remains of a prosperous past, which led the geographer Ellsworth Huntington (1911) to assert that in ancient times Transjordan had been endowed with a better climate, abundant water, and fertile soil. However, not until about half a century later did scholars begin to readdress the issue of ancient landscapes and climate change by engaging in systematic research to obtain data for better understanding the environmental history of this country.

By the turn of the twenty-first century, almost a century after Huntington wrote his ideas on the Transjordanian landscape, researchers had produced a considerable amount of information appropriate for reconstructing the ancient climates and landscapes of Transjordan. To a large extent, data have been obtained from a variety of sources and by a wide spectrum of sciences. The present volume focuses mainly on geoarchaeological and cultural ecological aspects of this research, supplemented by physical, chemical, and biological data obtained by a variety of scientists.

Geoarchaeological Approach

The evidence for reconstructing past environments is found in a variety of deposits, both inside and outside archaeological sites. Such deposits contain macro- and microscopic evidence that leads to reconstructing vegetation, fauna, climate, geomorphological processes, and the effects of human agency on the environment. To interpret the paleoenvironmental evidence extracted from sedimentary deposits, geoarchaeologists use information obtained through empirical methods on modern environments. For example, the size, shape, and arrangement of sediments in modern rivers are used as parameters in reconstructing past fluvial processes.

Although a young scientific approach, geoarchaeology has developed different areas of specialization derived from the diversity and complexity of geomorphic environments and the advances in science and technology. Consequently, views regarding the foundations of geoarchaeology vary. Some researchers see geoarchaeology within the scope of geomorphic processes, soils, and sediments (e.g., Holliday 1992; Waters 1992; Rapp and Hill 1998; Goldberg, Holliday, and Ferring 2001; Stein and Farrand 2001), while others see it in the application of geological, geochemical, and geophysical techniques (e.g., Pollard 1999; Goldberg, Holliday, and Ferring 2001). Other approaches within geoarchaeology put emphasis on human ecological aspects (e.g., Butzer 1982; Wilkinson 2003). Whatever the approach, geoarchaeology has become an integral part of most archaeological projects, whether survey or excavation.

The number of geoarchaeological studies in the eastern Mediterranean and Near Eastern regions varies considerably from country to country. The relatively large number of studies in Greece, Turkey, Israel, and Egypt contrasts with the relatively low amount of research in the rest of the countries in the region. Continuous political instability during the last quarter of the twentieth century discouraged studies in various Middle Eastern countries. Civil war in Lebanon, Syria's political isolation, the Arab-Israeli conflict, the Iranian Revolution, the Iran-Iraq War, economic embargoes on Iraq and Libya, and the Gulf wars of 1991 and 2003 reduced the opportunities for foreign and local scholars to pursue field research. However, even as political tensions increased in the region, the Jordanian authorities encouraged domestic and foreign teams to carry out archaeological research in the country to promote the existence of Jordanian antiquities abroad and to attract tourists. Thus, archaeological investigations in Jordan during the preceding three decades increased considerably. The number of articles containing geoarchaeological research in the *Annals of the Department of Antiquities of Jordan* in the first decade of the millennium have doubled, compared to the early 1980s. Similarly, in the past ten years, more articles on geoarchaeological research in Jordan have appeared in earth science and environmental journals such as *Geoarchaeology* and *The Holocene*.

Originally, the application of geoarchaeology was more popular in prehistoric archaeology projects, probably because geomorphology was essential in dating prehistoric sites, assessing the degree of disturbance by deflation and other forms of erosion, and placing the sites in the context of natural resources and paleoclimates. However, more recently, geoarchaeological research projects have also become popular in the archaeology of historical periods, where there is a need for understanding the relations between landscapes and complex societies.

Although often seen as specialists of geology, geomorphology, and soils, most geoarchaeologists make contributions to the reconstruction of human-environmental research. That is to say, geoarchaeologists look at the human dimension of the natural processes they study. In the Levant, geoarchaeology has contributed to the understanding of a variety of issues, such as climate change, the impact of human activities on the landscape, natural disasters in the past, and the dating of natural and cultural events. The reconstruction of natural and cultural events in the ancient past requires the integration of physico-geographical data (i.e., geological, geomorphological, biogeographical) with cultural data recovered by archaeologists. Therefore, in addition to possessing a deep knowledge of geomorphic processes, sedimentology, and soil formation processes, the geoarchaeologist must also understand how humans interact with their environment.

Geomorphological assessment studies in archaeology are essential not only for placing archaeological materials in spatial context but also for reconstructing landscape taphonomy (*sensu* Wilkinson 2003), which is a systematic study of the history of deposition and postdepositional disturbances that mask or destroy archaeological and paleoecological evidence. For the most part, it is the job of the geoarchaeologist to reconstruct the taphonomy of the landscape to put together the pieces of the puzzle that show what the landscape looked like at some point in the past. In an environment such as Jordan, where erosion, floods, earthquakes, and humans have taken a toll on prehistoric and even historic sites, the necessity for a geomorphological assessment is obvious.

Cultural Ecological Approach

Karl Butzer's (1982) *Archaeology as Human Ecology* had a sound impact on both geoarchaeology and cultural ecology, as it addressed not only aspects of earth and life sciences in archaeology but also interpretation of cultural patterns in the landscape. This approach led the new generation of geoarchaeologists to seek support in cultural ecology as the bridge to link the physical and societal environments.

Embedded within the broader field of cultural geography, cultural ecology is a research approach deeply grounded on paradigms developed in geography and anthropology. Its main purpose is to study the interrelationships between people, resources, and space (Butzer 1989). Thus, cultural ecology is the basis for understanding processes such as adaptation, adaptive strategies, environmental perception, and ways in which cultural groups interact with their environment. In the New World, the cultural ecological approach sparked a new line of research in the early 1990s, on the occasion of the quincentenary of Columbus's discovery of the Americas (Butzer 1992).

In the Mediterranean region, cultural ecological approaches in landscape reconstruction have a longer tradition. *Beyond the Acropolis* (Van

Andel and Runnels 1987) is one of the studies that sparked interest in paleoenvironmental research with a strong cultural component. Subsequently, the publication of *Man's Role in the Shaping of the Eastern Mediterranean Landscape* (Bottema, Entjes-Nieborg, and van Zeist 1990) marked the beginning of a new wave of studies focusing on the importance of the cultural aspects in the reconstruction of Near Eastern paleoenvironments. Thus, modern paleoenvironmental research in the Near East addresses several cultural ecological problems, of which the most popular are (1) environmental factors in the rise and collapse of ancient civilizations, (2) landscape deterioration, and (3) prehistoric adaptations.

Although the topic of rise and collapse of civilization varies from region to region, one of the relevant advances regarding this subject during the 1990s is the interest in the environmental causes of civilization collapse at the end of the third millennium BC. *L'Urbanisation de la Palestine à l'Âge du Bronze Ancien* (De Miroschedji 1989) prompted the interest in and discussion of the possible causes of collapse, which are further discussed in *Third Millennium BC Climate Change and Old World Collapse* (Dalfes et al. 1997). Although large amounts of research have addressed the environmental crisis and societal collapse at the end of the Early Bronze Age, the causes for the collapse remain a controversial issue among Near Eastern archaeologists.

Environmental degradation, another controversial issue, is often addressed through a variety of studies dealing with paleoenvironmental reconstruction. The initial issue in assessing landscape degradation is examining cultural aspects attached to the term *degradation* in relation to deterioration and transformation of the landscape. Therefore, these terms (as used throughout the book) are defined here.

Environmental degradation does not necessarily imply full irreversible destruction of the landscape. Piers Blaikie and Harold Brookfield (1987) present the term *net degradation* as a result of natural and human forces shaping the landscape. Within each force, there are destructive and reconstructive directions that interact as defined in the following equation: net degradation = (natural degradation process + human interference) − (natural reproduction + restorative management) (Blaikie and Brookfield 1987, 7). This equation implies that if the natural recovery of the landscape and the care taken during its management are greater than the natural and human destructive processes, then the net degradation of the landscape will result in a relatively healthy landscape. Problems arise, however, when landscape resilience and restoration efforts are smaller than the rate of destruction.

Deterioration is not necessarily a cultural term, for deterioration can result from climate change, geomorphic processes, and other natural processes. Thus, the effects of the Younger Dryas on the landscapes of the Near East, often recognized as aridization, would be a form of landscape deterioration that is totally independent from human action.

William Beinart and Peter Coates (1995) acknowledge the word *transformation* as indicating a somewhat longer-term modification of the landscape and also serving as a less emotive concept than degradation. This definition suggests that degradation spans longer periods of time, through which resilience of the landscape is implied. Oftentimes, however, the term *degradation* is limited to a certain aspect of the environment, such as *soil degradation*. In this book the terms *landscape degradation* and *environmental degradation* are used to encompass multiple aspects (soil, vegetation, fauna, streams, lakes, coasts, etc.).

The main forms of environmental degradation often mentioned in studies of landscape in the ancient Near East include deforestation and impoverishment of vegetation, reduction of wildlife, slope destabilization and soil erosion, transformation of streams and water bodies, salinization, and pollution of soil, water, and air (Cordova 2005a). Fossil pollen sequences are the most common source of data for reconstructing vegetation history and assessing deforestation and landscape disturbance through agricultural activities. Sequences of alluvial deposits are the most common way to reconstruct the history of streams and a way to assess upland soil erosion. The compilation by Sytze Bottema, Gertie Entjes-Nieborg, and Willem van Zeist (1990) contains various studies on different forms of past environmental degradation. Most recently, Tony Wilkinson (2003) has analyzed the millennial process of land degradation in the Near East by pointing to details still visible in the landscape.

The cultural ecological topics concerning Levantine prehistory are diverse because of the long time span involved (1.8 million years) and the variety of aspects (resources, migrations, diffusion of techniques, and adaptation to climate change). One of the main topics in the Lower and Middle Paleolithic is the early hominid migration to Eurasia. *Neandertals and Modern Humans in Western Asia* (Akazawa, Aoki, and Bar-Yosef 1998) and *Neanderthals in the Levant: Behavioral Organization and the Beginnings of Human Modernity* (Henry 2003a) contain the latest advances in a variety of topics within this broad subject, subsequently stirring new research trends. For instance, the alleged coexistence of Neandertals with modern humans during the transition from Middle to Upper Paleolithic is one of the issues drawing increasing attention.

Another cultural ecological focus of prehistoric research in the Levant is climate and landscape change during the Pleistocene-Holocene transition and their influence on the origins of farming and pastoralism. Compilations of research that have been key to this subject include *The Natufian Culture in the Levant* (Bar-Yosef and Valla 1991) and *The Origins and Spread of Agriculture and Pastoralism in Eurasia* (Harris 1996). The most comprehensive analysis of the environmental and cultural aspects leading to the development of agriculture in the southern Levant is *From Foraging to Agriculture: The Levant at the End of the Ice Age* (Henry 1989), a book that had a tremendous impact on subsequent research in this topic.

Additionally, compilations of works dealing with advances in prehistoric research in Jordan include Garrard and Gebel 1988, Gebel, Kafafi, and Rollefson 1997, Henry 1998a, and MacDonald, Adams, and Bienkowski 2001, all of which stress research in the areas of Quaternary research, geoarchaeology, and human-environment relationships.

Because most of the research topics mentioned above result from geoarchaeological research and often deal with adaptation, migration, diffusion, and other aspects of cultural ecology, the discussion of data in this book falls within the scope of geoarchaeology and cultural ecology.

Landscapes and Paleolandscapes

The concept of landscape is crucial to the scope of geoarchaeology and cultural ecology because it is the basis for visualizing the stages of environmental change at a particular locale. Although geographers and archaeologists use this concept for interpretations of different time frames, the concept is essential to both disciplines. Many of the concepts of landscape in these geographical disciplines have a mirror in the conceptualization of archaeological landscape, which according to Wilkinson (2003) includes social, economic, and physical aspects. This means that landscapes are made of not only soils, plants, and animals but also features such as dams, buildings, and roads. Cases of purely natural landscapes are virtually nonexistent in the scope of archaeological and geographical research. In cases such as the islands of New Zealand, where human occupation is relatively recent (approximately one thousand years), the study of natural landscapes becomes of importance for understanding the impact of human-induced environmental change. However, this is evidently not the case in the Near East, where human agency has been present during hundreds of thousands of years.

The paleolandscape is the best way to visualize environmental change in the past after sets of paleoecological data are put together. The process of landscape reconstruction starts with the understanding of the modern landscape. The second stage is the process of sampling, processing, and obtaining data to build records of paleoenvironmental proxy data. Data sets of pollen, diatoms, and the isotopic composition of organisms in a lake bed are difficult for the nonspecialist to understand. The same can be said for a series of soil profiles and sequences of alluvial sediments, which could mean nothing to a layperson if they are not first interpreted by a specialist. For example, a black organic soil buried under one or two meters of younger sediment could mean many things. However, organic matter content and the presence of grass pollen could be clues that this soil formed under conditions of prairie vegetation. Based on this information, the final interpretation is that at a certain point this location was a landscape dominated by grassland, which is easily understood because we know what grasslands look like and what kind of resources they may have offered to the people who lived on them. Analogs in the modern landscapes can thus be used in the interpretation of sets of paleoecological proxy data. Finally, each component of the landscape (e.g., vegetation, soils, climates) is figured out before this information is integrated in the final step of the process, that is to say, defining what the given landscape looked like at a certain time in the past.

The brief overview of these steps is a simplistic way to show how specialists of different disciplines interpret their data to reconstruct the components of paleolandscapes, but it conveys the idea that interpretation of the paleolandscape relies on an understanding of present landscapes. This is why the interpretation of modern landscapes, which is one of the aspects this book emphasizes, is important.

Physical and Cultural Diversity

Four of the five Middle Eastern floristic provinces defined by Michael Zohary (1973) are present in Jordan (figs. P.1 and 1.1). This results in an enormous diversity of vegetation, climate, geology, landforms, and soils within the relatively small territory of Jordan.

Because vegetation is one of the most distinctive visible aspects of the landscape, most of the geographical interpretation presented in this book focuses on the four vegetation provinces of Jordan: Mediterranean, Irano-Turanian steppe, Saharo-Arabian desert, and Sudanian desert (fig. P.1, preface). Although they are chiefly defined by their characteristic floras, each of these regions presents distinctive ecological characteristics (soil, fauna, land use, etc.). Therefore, in this book they are referred to as "landscape regions." Each landscape region represents specific challenges for adaptation, and these adaptations result in cultural traits particular to each region. The Mediterranean region of Jordan has been the typical rain-fed agricultural area, with strong ties to the greater Mediterranean realm. Olive cultivation, winter cereals, grapevines, gardening, and a combination of farming and pastoralism are the main aspects of rural life in this part of Jordan. In contrast, the Irano-Turanian steppe is an area where rain-fed farming is highly unreliable. Therefore, the traditional rural economy there is based on pastoralism, which is a mobile activity with links to the Mediterranean region and the desert. The deserts have traditionally been winter or early spring pastoral grounds. As edible herbs disappear, the pastoral activity migrates to the Irano-Turanian steppe and eventually into the Mediterranean region. Although rain-fed agriculture is not possible in the desert, cultural strategies have enabled farming in some areas where flood irrigation is feasible. Another strategy includes the farming of plants resistant to drought, a strategy that incorporates planting in the less dry grounds and gambling on sporadic rains, as practiced by modern Bedouins in the drylands of southern Jordan (Henry et al. 2003).

Rural management strategies evolve as the landscape changes. In some cases, archaeological evidence in a site in the steppe may reflect rural activities carried out within a woodland or forest, which is an interesting piece of information for paleoenvironmental reconstruction. For example, plant and animal remains recovered from Pre-Pottery Neolithic B levels in Ain Ghazal indicate a wooded environment between ca. 9 ka and 8 ka BP (Köhler-Rollefson and Rollefson 1990). However, the site is today located in a largely treeless landscape, within the region referred to by Dawud al-Eisawi (1985) as the "non-forest Mediterranean region." The same situation occurs in the Pre-Pottery Neolithic settlement of Ain Abu-Nukhaylah in the sandstone desert of southern Jordan, where archaeological, zooarchaeological, and archaeobotanical evidence suggest a steppe rather than a desert environment during its occupation around 8 ka BP (Henry et al. 2003). Many of the plant species do not exist in the region today but occur farther south, in the Arabian Peninsula, where summer rains allow a different flora. The evidence points to both higher moisture and incidence of summer rains. Proper knowledge of modern local and regional flora and fauna is therefore necessary for interpreting assemblages of plant and animal remains found in cultural deposits. For example, understanding the distribution of modern physical and cultural components of woodland remains in the modern landscape (chapter 3) is essential for understanding basic ecological aspects of prehistoric woodland ecosystems. The same can be said for steppes and deserts, which have been profoundly transformed, making them different from the ones prehistoric peoples inhabited (chapter 4).

The transitional character of most landscapes makes the tracing of their boundaries difficult, because changes between landscape zones are not sharp but gradual. Often referred to as ecotones, these transitional landscape boundaries are ecologically and culturally rich because characteristics of both sides mingle. Therefore, recognizing what types of flora and fauna are adapted to these ecotones is important. Two landscape regions share plant species — as is the case, for example, with a large number of herbaceous plants in the Mediterranean region and the Irano-Turanian steppe — but only one set of plants within this group is adapted to the changing nature of the ecotones. The knowledge of these transitional plants may shed light on the interpretation of archaeobotanical and palynological data in the reconstruction of boundary fluctuations. Soils are another source of information for reconstructing landscape change. Climatic and geomorphic factors leave their imprint on soils through the development of horizons. If well preserved and properly interpreted, soil profiles may provide information that is significant in reconstructing fluctuations of landscape boundaries.

The Role of Modern Landscapes in Paleoenvironmental Reconstruction

The scarcity of paleoenvironmental data in Jordan makes the reconstruction of landscapes difficult, especially when most evidence presents time gaps. However, flora, fauna, and soils in modern landscapes may be used as reference points in the interpretation of archaeobotanical, archaeozoological, and geoarchaeological data. Nevertheless, one should consider that the landscapes of Jordan have been transformed to such a degree that determining what the preagricultural landscapes looked like is difficult. Therefore, it is important to assess the stages of landscape transformation during the millennia since the first farming communities appeared, and for that purpose, the modern woodlands, steppes, and deserts of Jordan still hold useful clues for understanding millennial landscape change.

2
The Physical Scene

Basic Physico-Geographic Facts

The territory of the Hashemite Kingdom of Jordan is 89,213 square kilometers (34,436 square miles), similar in size to Portugal, Hungary, Azerbaijan, and French Guyana. Jordan's territory is slightly larger than the state of South Carolina and slightly smaller than the state of Indiana. Jordan's northern and southern latitudes are 33°23′ N and 29°10′ N, respectively. Its extreme western and eastern longitudes are 34°57′ E and 39°25′ E, respectively. Cities around the world located within this latitude range include Marrakech, Morocco; Tripoli, Libya; Cairo, Egypt; Tel Aviv, Israel; Basra, Iraq; Shiraz, Iran; Lahore, Pakistan; Amritsar, India; Shanghai, China; Nagasaki, Japan; San Diego, California; Tucson, Arizona; Ciudad Juarez and Tijuana, Mexico; Austin, Texas; New Orleans, Louisiana; and Jacksonville, Florida.

Jordan has common borders with Israel and the Palestinian West Bank to the west, with Syria to the north, with Saudi Arabia to the south and east, and with Iraq to the northeast. Jordan's sea shoreline is only 26 kilometers (16.16 miles), at the Gulf of Aqaba on the Red Sea.

The highest elevation is located at 1,754 meters (5,755 feet) above sea level, on the summit of Jebel Ram; the lowest is at 408 meters (−1,339 feet) below sea level, on the surface of the Dead Sea. Five of the six largest Jordanian cities (Greater Amman, Zarqa', Irbid, Salt, and Madaba) are located at elevations between 700 and 1,100 meters (2,297–3,609 feet), which comprises those areas with the Mediterranean type of climate. Aqaba is located only a few meters above the Red Sea, where the city is exposed to extremely high temperatures during the summer. However, Aqaba is Jordan's only seaport and the only outlet to the world oceans. Approximately 20 percent of the country lies below sea level: these areas include the Jordan Valley, the Dead Sea basin, and the lower valley of the Wadi Araba.

Jordan's standard time is two hours ahead of Greenwich mean time, which makes it the same as Finland, Estonia, Latvia, Lithuania, Bela-

rus, Ukraine, Turkey, Syria, Lebanon, Israel and the Palestinian territories, Egypt, Sudan, Uganda, Zambia, Zimbabwe, Mozambique, and
South Africa, among other countries. This means that Jordan is one hour
ahead of Paris, two hours ahead of London, seven hours ahead of New
York and Toronto, eight hours ahead of Chicago and Dallas, nine hours
ahead of Denver and Calgary, and ten hours ahead of Los Angeles and
Vancouver.

The three major water courses are the Jordan River (a common
boundary with Israel and the West Bank), the Yarmouk River (which
marks a segment of border with Syria and the Golan Heights), and Wadi
Zarqa', which is a tributary of the Jordan River with its basin inside the
territory of Jordan. The Dead Sea (shared with Israel and the West Bank)
is the major body of water.

Tectonics and Surface Geology

Jordan and the Regional Tectonic Context

The location of Jordan at the convergence of three tectonic plates plays an
important role in the distribution of rock formations and geologic structures in the country. The limits between the Arabian Plate, the African
Plate, and the Levantine Plate are marked by a north–south rift system
formed by the Dead Sea–Gulf of Aqaba Rift and the Suez Rift (fig. 2.1).
Tectonic activity along this rift system possibly began in the Precambrian, affecting the crystalline rocks that form the geological basement
of Jordan (Bender 1974).

At the beginning of the Miocene, a bifurcation in the Red Sea Rift
created a north–south graben that separated the Levantine Plate from
the Arabian Plate (Meissner 1986). This graben is the precursor to the
rift depression that is now occupied by the Jordan River, the Dead Sea,
and the Wadi Araba. Post-Miocene movements resulted from the development of major faults paralleling the down-thrown block of the graben
(Garfunkel 1997). Although the majority of faults are strike-slip, more
recent tectonic movements are characterized by normal faults (Neev and
Emery 1995).

East–west-trending faults cut across the Rift Valley, producing dis-
locations in the bottom of the rift and creating uplifted blocks. One of
these faults marked the division of the Dead Sea basin into two sub-
basins, north and south, which are divided by the uplifted block that

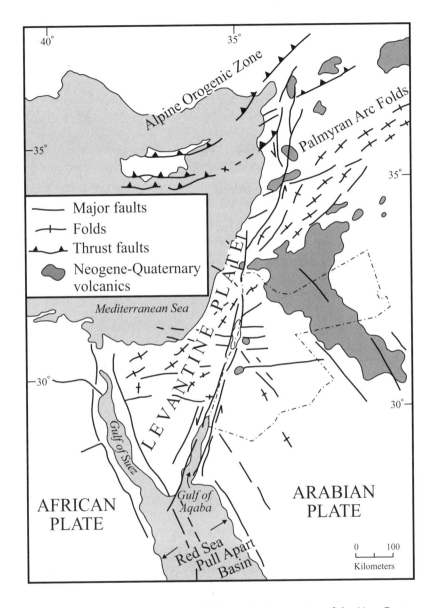

Figure 2.1 Jordan in the context of the regional tectonics of the Near East.

forms the Lisan Peninsula (Neev and Emery 1995). During the Pliocene, accumulation of salt in the sediments of the Dead Sea basin created upward movements, thus giving rise to salt diapirs (Zak 1997).

Plate movements and stresses within the earth's crust produced a complex system of faults running perpendicular or diagonal to the rift (fig. 2.1). This fault system eventually controlled the drainage network that dissects the plateaus of western Jordan, as they captured streams that once ran east, bringing these courses into the Rift Valley (Beheiry 1968–69; Al-Hunjul 1995; Shawabekeh 1998). Some of these faults were also involved in the development of volcanism that produced the basaltic flows on the Karak Plateau and in the Wadi Zarqa' Ma'in and created thermal springs such as those of the Wadi Zarqa' Ma'in (Bender 1974; Steinitz and Bartov 1992). Faults associated with the stresses in the Rift Valley also extended into the eastern part of the Jordanian plateau, where they contributed to the formation of depressions such as Azraq and Al-Jafr and to volcanic activity in the northwestern plateau (Beheiry 1968–69; Bender 1974).

Although remains of faulting, volcanism, and hydrothermal activity are found all over Jordan, the intensity of these phenomena has been concentrated in the Rift Valley, where intense tectonic activity continues to this day. The best expression of current tectonic activity rests on the frequency of earthquakes and active faulting. Earthquakes have been responsible for the destruction of numerous prehistoric and historic settlements along the rift (Neev and Emery 1995; Niemi, Ben-Avraham, and Gat 1997; Enzel, Kadan, and Eyal 2000).

Major Geologic Units

Jordan's geological surveys date back to the British mandate period, but not until after independence was the geology of the kingdom defined in more detail. David Burdon (1959) published a chronology of Jordan's geology, which later was modified and worked out in greater detail, mainly through Friedrich Bender's geological survey during the 1950s and 1960s. This information was made available in a compendium published first in German (Bender 1968) and later in English (Bender 1974) and in a series of maps. Bender's geological survey of Jordan became a fundamental source of information for all subsequent geological studies.

Today, more detailed geological monographs of local and regional detail exist for most regions of the country. An important source of information exists in the bulletins accompanying the 1:50,000 geological maps published by the Natural Resource Authority (NRA). In addition, numerous scientific journal articles and theses provide relevant information on particular aspects of the geology of Jordan. In sum, the amount of geological information obtained and published so far in Jordan is vast. Not all of the details concerning geological units in Jordan can be mentioned in this chapter, but for the purpose of a general description, the most important aspects of Jordan's Pre-Quaternary geology are summarized and grouped into six major unconventional units (fig. 2.2).

The general sequence of rocks from oldest to youngest can be appreciated along a southwest–northeast transect across Jordan from Aqaba to the Syrian border (transect C–C', fig. 2.2). The oldest rocks belong to the crystalline Precambrian Basement and are exposed in the southwestern corner of the country, mainly in the southern Wadi Araba and areas around Aqaba. The youngest rocks correspond to the Plio-Pleistocene basalts of the Badia region in the northeastern part of the country. In between, there are dozens of sedimentary formations of terrestrial and marine origin spanning from the Precambrian to the Quaternary. Transects from the Rift Valley to the highlands also show massive packages of rock formations and the intense faulting associated with the tectonic development of the Rift Valley (transects A–A' and B–B', fig. 2.2).

The ancient basement rocks of the Aqaba region (which include the Precambrian Basement complex) are exposed along the Wadi Araba, the Wadi al-Yutm basin, and the base of mountains in the Wadi Rum area. These rocks are part of the Nubian-Arabian Shield that extends to the northeast from the Sinai and the eastern side of the Gulf of Aqaba (Bender 1974). Northward and northeastward, the formations of the Nubian-Arabian Shield disappear under younger rock formations. The main rocks of the basement belong to the intrusive igneous types such as granite, granodiorite, and porphyry, as well as some metamorphic rocks such as gneiss.

The sandstone formations of the Rift Valley and Wadi Rum are grouped into two broad groups — the Ram and the Kurnub Sandstones — both of which dominate in southern Jordan and along the escarpment facing the Rift Valley. These are the rocks that form the impressive topography of Wadi Rum and Petra.

The limestone and other marine rocks of the Transjordanian plateau correspond to sedimentary formations of Upper Cretaceous and Eocene age that constitute the bulk of the rocks forming the Western Highlands

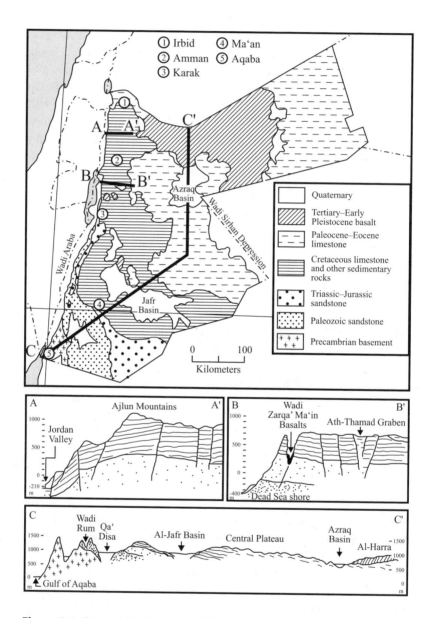

Figure 2.2 General distributions of lithological groups and transects across Jordan.

and the Central Plateau (fig. 2.2). This group makes up about half of the surface geology of Jordan. Although limestones are the main type of rock, these formations contain a variety of other sedimentary rocks, such as marls, travertines, and phosphorites, among others. The strata of the formations comprising these series lie horizontally and subhorizontally, consequently influencing the development of the plateau morphology that characterizes some parts of the Western Highlands (fig. 2.2, transect B–B′). Intense faulting and deformation of strata have formed topographic features such as the Ajlun Mountains (fig. 2.2, transect A–A′). Eocene sedimentary rocks such as the Umm Rijam Chert Limestone are frequently found filling in early Tertiary depressions such as the Jafr basin and the Thamad graben. The Cretaceous and Eocene limestone formations are the main sources of flint in Transjordan, in particular the Amman Silicified Limestone (Quintero 1996) and the Eocene Umm Rijam Chert Limestone (Quintero, Wilke, and Rollefson 2002).

During most of the Tertiary period, terrestrial deposits accumulated in the Rift Valley as the bottom of the graben sank. Terrestrial sedimentary units of Tertiary age exposed in the rift consist mainly of conglomerates and breccias, some of which have been dislodged by faulting, intruded by younger basalts, and severely eroded.

The Dana Conglomerate of Miocene age is found in several spots along the rift. Although the accumulation of these conglomerate units preceded the advent of hominids in the Near East, they were an important source of lithic material for Lower and Middle Paleolithic peoples. Lower Paleolithic (Acheulian) tools are often found on the deposits of the Dana Conglomerates in the rift (Macumber and Edwards 1997). The clasts forming these conglomerates derived from the erosion of Cretaceous and Eocene chert-rich limestones. Therefore, they contain numerous boulders and cobbles of chert, which were used as reduction cores.

Tertiary volcanic rocks in Jordan are concentrated in two locations: along the Rift Valley and in the northeastern plateau. Smaller isolated basalt units appear in some areas of the plateau, such as the region between Wadi al-Hasa and Al-Jafr basin, where volcanic activity is associated with fault systems formed perpendicular to the Rift Valley. Potassium argon (K-Ar) dates from the Rift Valley span from 24.8 to 0.5 Ma (Steinitz and Bartov 1992). In northeastern Jordan, K-Ar dates span from 8.9 to 0.1 Ma (Allison et al. 2000).

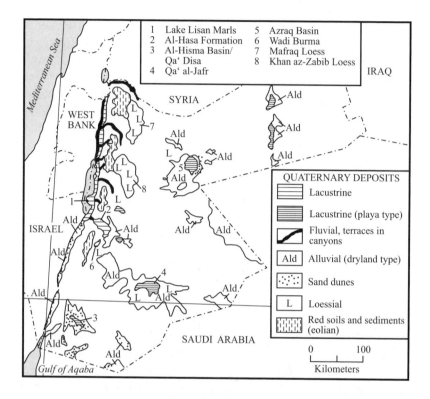

1	Lake Lisan Marls	5	Azraq Basin
2	Al-Hasa Formation	6	Wadi Burma
3	Al-Hisma Basin/	7	Mafraq Loess
	Qaʿ Disa	8	Khan az-Zabib Loess
4	Qaʿ al-Jafr		

QUATERNARY DEPOSITS

Lacustrine
Lacustrine (playa type)
Fluvial, terraces in canyons
Ald	Alluvial (dryland type)
Sand dunes
L	Loessial
Red soils and sediments (eolian)

Figure 2.3 General distributions of Quaternary rocks and deposits across Jordan.

Quaternary Geology

The Quaternary system corresponds to approximately the past 1.8 Ma, an insignificant part of the entire geological history. However, the Quaternary is the most relevant to archaeologists because humans developed during this time. In addition, Quaternary deposits (fig. 2.3) are a source of proxy data for paleoenvironmental reconstruction.

Basalts of Quaternary Age

Although most of the volcanic activity in Jordan took place during the Miocene and Pliocene periods, minor activity occurred in the Lower Pleistocene. The Ghor al-Katar basalts are overlain by the Lisan beds and underlain by a conglomeratic deposit bearing early Acheulian lithic materials (Bender 1974, 94, 100). Therefore, this association would place the age of basalts somewhere in the Lower Pleistocene (1.8 Ma–750 ka BP) or, at the latest, in the Middle Pleistocene (750–125 ka BP).

Absolute dates for basalts are few but confirm their Plio-Pleistocene age. Investigations using K-Ar dating and associated lithic materials have confirmed the early Pleistocene age of some basalts in the Rift Valley and northeastern Jordan. K-Ar age determinations from the β-Basalt in the Wadi Zarqa' Ma'in produced dates spanning the period 1.8–0.5 Ma (Steinitz and Bartov 1992). Many other undated basalts in the Rift Valley occupy a similar stratigraphic position, which suggests that Plio-Pleistocene volcanism occurred in several locations. On the Northern Basalt Plateau (Al-Harra), K-Ar dates also confirm the occurrence of Lower Pleistocene volcanism that formed scoria cones (Al-Malabeh 1994; Allison et al. 2000).

Lacustrine and Fluvio-Lacustrine Deposits of the Rift Valley

The Quaternary sediments deposited in the lacustrine basin of the Rift Valley are exposed in numerous locations along the margins of the Jordan Valley, the Dead Sea basin, and the northern part of the Wadi Araba. Based on a large body of previous studies, a comprehensive summary of post-Miocene stratigraphic units of the western side of the Dead Sea Rift has been produced (Stein 2001). In general terms, the stratigraphic sequence includes the Sedom, Samra (Amora), Lisan, and Ze'elim formations.

The Sedom Formation comprises a sequence of late Pliocene to early Pleistocene salt and gypsum deposits. The depositional environment of the Sedom Formation consists of brines formed in a lagoon that was probably connected to the Mediterranean Sea (Zak 1997). Calcium and chloride contained in the Sedom brines provided the ingredients for the formation of aragonite and gypsum in younger deposits (Zak 1997; Stein et al. 2000).

The Samra Formation (also known as the Amora Formation) consists of marls, salt, chalk, and sandstone beds deposited at the bottom of former Lake Samra (Lake Amora) (Stein 2001). Uranium series dates obtained from deposits near the Mt. Sedom area in Israel (Kaufman, Yechieli, and Gardosh 1992) put the upper part of the Samra Formation between 200,000 and 80,000 years ago. The calcium carbonate phase in the lower part of the formation is aragonite, while that of the upper one is calcite. The transition from aragonite to calcite implies an increase in freshwater supply (Stein 2001). Despite the knowledge of stratigraphy and mineralogy of the Samra Formation, the extension and maximum levels of former Lake Samra are not known.

The Lisan Formation overlies the Samra Formation on a stratigraphic unconformity, which could imply a depositional hiatus (Stein 2001). This means that the existence of the two lakes was not consecutive, but separated by a period of no deposition. The Lisan Formation has been subdivided into three members, recognized on the east and west margins of the Rift Valley. On the western margin of the Dead Sea, at Perazim Valley and Massada Plain, the Lower Member and Upper Member are characterized by packages of alternating laminae of aragonite and detritus, deposited under rising lake-level conditions (Bartov et al. 2002). The Middle Member is subdivided into two parts divided by a depositional hiatus, which in turn marks a change from a higher to a lower sedimentation rate that coincides with the rapid rise of the lake level between circa 30 ka and 26 ka BP (Bartov et al. 2002).

On the Jordanian side of the Rift Valley, the Lisan Formation has been studied in the area around Damya and Karamah, in the Jordan Valley, and in the Lisan Peninsula (Abed and Helmdach 1981; Abed and Yaghan 2000; Landmann et al. 2002). In the Damya area, a Holocene deposit of sands overlying the Lisan Formation was named the Damya Formation, which is often equated with the Ze'elim Formation of the western margin (Landmann et al. 2002). The Lisan Formation on the Jordanian side mirrors the members and facies studied on the west side of the rift. Accordingly, a Lower, Middle, and Upper Member have been distinguished, corresponding to lacustrine accumulation between 65 ka and 15–16 ka BP. The top of the laminates of the Upper Member ends with the White Cliff aragonite varves, which correspond to the driest periods of the Last Glacial Maximum (Abed and Yaghan 2000). Overlying the White Cliff varves is the Damya Formation, which contains more lithogenic material, suggesting a gradual decrease of lake levels and an increase in terrigenous sediments (Landmann et al. 2002).

The decline in Lake Lisan levels began after 17 ka BP and reached its minimum level between 12 ka and 11 ka BP. This minimum is estimated to have been about 725 meters below sea level, based on reflection data from the bottom of the Dead Sea (Neev and Hall 1979). This low level means not only the end of Lake Lisan but also the near disappearance of the body of water in that location. The lake level rose again starting circa 10 ka BP, thus marking the beginning of the Holocene Dead Sea, whose levels have fluctuated between −300 meters and −500 meters (Frumkin et al. 1994; Stein 2001). The Ze'elim Formation, which forms the top of the lacustrine sequence, corresponds to the Early Holocene Dead Sea lacustrine deposits (Yechieli et al. 1993).

The fluvio-lacustrine deposits of the Dead Sea Rift include a variety of sedimentary units of fan-deltas, some of which have provided stratigraphic information on Pleistocene lake-level change (Machlus et al. 2000) and Holocene earthquakes (Enzel, Kadan, and Eyal 2000). In the areas of Tabaqat Fahl and Wadi Hammeh, geomorphological and stratigraphic studies have shown a sequence of lake and delta fluctuations in relation to climate change, tectonics, and human occupation during the Terminal Pleistocene. The Hammeh deposits, known as the Knob Limestone, correspond to fluvio-lacustrine facies that formed as the waters of Lake Lisan receded (Macumber and Head 1991).

Lacustrine Deposits East of the Rift Valley

Most of the lacustrine deposits east of the Rift Valley are concentrated in the endorheic basins of the Central Plateau, where Qaʿ al-Azraq, Qaʿ al-Jinz, Qaʿ Disa, and Qaʿ al-Jafr are the largest basins (fig. 2.3). They contain thick deposits of lacustrine and eolian material of Pleistocene age. Clay layers obtained from deep cores of these deposits suggest that deep-water environments existed in the basins (Davies 2000, 2005). Qaʿ al-Jafr shows that not all deposition in the basin was purely lacustrine, as a mixture of eolian and alluvial deposits are present there (Davies 2005).

Lacustrine deposits are also found in numerous locations along the valleys dissecting the western part of the plateau, indicating that water was impounded in these valleys sometime during the Pleistocene. The largest of such basins corresponds to the upper reaches of the Wadi al-Hasa, where the remains of such lakes are recorded in the marls of the Hasa Formation.

The Hasa Formation (fig. 2.3) has been described and dated in several localities around Jurf ad-Darawish and Wadi Burma (Moumani, Alexander, and Bateman 2003). In addition, the upper members of this formation in the northwestern part of this region, near the area where the Wadi al-Hasa becomes entrenched in the plateau, have been described (Vita-

Finzi 1966; Copeland and Vita-Finzi 1978; Schuldenrein and Clark 1994, 2001). The Burma Member corresponds to lacustrine sediments formed under the presence of Lake Burma, which seems to have been connected with former Lake Hasa (Moumani, Alexander, and Bateman 2003).

The lacustrine facies of the Hasa Formation correspond to the Burma Member. Optically stimulated luminescence (OSL) dates put the lacustrine facies of the Burma member between 111 ka and 20 ka BP, although the duration of the lake may be extended, ranging from as early as 125 ka to about 19 ka BP (Moumani, Alexander, and Bateman 2003). This stratigraphic sequence is not continuously lacustrine, since conglomerate deposits embedded in it suggest a transition to a fluvial environment. Additionally, calcretes in the sequence suggest times of exposure to high evaporation rates.

Fluvial Deposits

This group of Quaternary sedimentary deposits includes a variety of fluvial depositional facies occupying terraces along the main valleys of the plateau and on the edges of the Rift Valley. Fluvial deposits can be roughly classified into conglomerates, alluvial fans, high terrace deposits, and alluvial valley fills.

Conglomerates of Quaternary age abound all along the escarpments bordering the Rift Valley. On the northwestern side of the rift, the Ubeidiya Formation is one of the best known because of numerous paleontological and geochronometric studies (Tchernov 1987; Bar-Yosef 1994). Similar conglomeratic deposits containing Acheulian lithics in Jordan include the Ghor al-Katar and Abu Habil formations (Abed 1985; Muheisen 1988; Macumber and Edwards 1997), the Kufrinja Gravels (Abed 1985), and the Tabaqat Fahl Formation (Macumber and Head 1991; Macumber and Edwards 1997). Numerous deposits of Pleistocene conglomerate form terraces in the lower reaches of the canyons emptying into the rift—such as Wadi Zarqa' Ma'in, Wadi al-Mujib, Wadi al-Karak, and Wadi al-Hasa—but such deposits have not been classified and dated.

One of the best-known conglomeratic deposits of Pleistocene age outside the Rift Valley is the hominid tool—bearing Dauqara Formation, which corresponds to the upper terrace in the Wadi Zarqa' (Baubron et al. 1985; Besançon and Hours 1985). In Wadi Burma, the conglomeratic facies of the Darwish Member of the Hasa Formation also contain

Lower Paleolithic stone artifacts (Moumani, Alexander, and Bateman 2003). Other unnamed conglomerates have been found in association with Paleolithic occupations in the Azraq basin, especially in Wadi Kharraneh, Wadi Uwaynid, Wadi Ratam, and Wadi Enoqiya (Copeland and Hours 1988; Besançon, Geyer, and Sanlaville 1989).

Although Pleistocene conglomerate deposits are plentiful in Jordan, they have received little attention. On most geological maps, they appear with designations such as "Pl" or "Plg," which stand for either fluviatile gravels or fluviatile plateau gravels. However, as the above-mentioned examples show, these deposits are often important for Lower and Middle Paleolithic research, since they contain faunal remains and evidence of tool making.

Alluvial fan deposits are located in southern Jordan, mainly along the Wadi Araba and along Wadi Rum in the south, and in the eastern desert areas, particularly around the large playa lakes. Alluvial fan deposits are varied and difficult to date. In part this is because of the erratic and ever-changing regime of deposition that characterizes these depositional environments. Nonetheless, various methods involving geomorphological, pedogenic, and archaeological criteria are used for dating alluvial fans. Despite these problems, three major phases of Late Quaternary alluvial fan development in the southeastern part of Wadi Araba have been recognized (Niemi and Smith 1999). Besides this local example, no regional chronology of alluvial fan deposition exists elsewhere in Jordan.

Holocene alluvial deposits in Jordan are better known but sparsely dated. Numerous Holocene alluvial units have been reported along the Jordan Valley (Mabry 1992) and the margins of the southeastern shore of the Dead Sea (Donahue 1981, 1984; Donahue, Peer, and Schaub 1997). A series of complicated units has been reported in the Wadi Faynan (Hunt et al. 2004). In the valleys dissecting the Western Highlands, Terminal Pleistocene and Holocene deposits form sequences of terraces. The most conspicuous are those associated with Natufian sites, which include, for instance, a series of terraces in Wadi Ziqlab (Field and Banning 1998), the Thamad Terrace in Wadi ath-Thamad (Cordova et al. 2005), the Tabaqa Terrace in Wadi Ahmar (Olzsewski and Hill 1997), and one of the high terraces in Wadi Judayd (Hassan 1995). The seriation of Holocene terraces is more complicated, since there is variation across the region. However, Middle Holocene terraces have been identified as the Intermediate Terrace in Wadi Ahmar (Hill 2002) and the Tur al-Abyad Terrace in Wadi ath-Thamad (Cordova et al. 2005), both of which are correlative with the alluvial unit designated as H alluvium (Mabry 1992) in the drainages of the eastern margin of the Jordan Valley.

Loess and Loess-Derived Deposits

Loess refers to silt deposits laid down by wind (primary loess), while loess-derived deposits are accumulations that originated from the re-deposition of primary loess by colluvial and fluvial processes (Pye and Sherwin 1999). The accumulation rates of primary loess are generally very low. Estimates from the Netivot Loess on the northern fringes of the Negev indicate a rate of 0.1 millimeters of deposition per year (Bruins and Yaalon 1979). Secondary loess deposits are located in depressions and valleys and are usually differentiated by mixtures of particles larger than silts. Thus, sand and gravel are common in loess-derived deposits.

Studies of loess deposits in Israel (Bruins and Yaalon 1979) and southern Syria (Rösner 1989) show that particle size, carbonate content, and trace minerals point to a nonglaciogenic origin. Thus, loess origin there is of the peridesert type, which occurs mostly from the deposition of dust that originated in the nearby deserts (Pye and Sherwin 1999). In the Levant, for example, dust particles are carried by storms from North Africa (Yaalon and Ganor 1979). Such events were evidently more frequent during the Pleistocene, since most loess deposition in the southern Levant predates the Holocene (Bruins and Yaalon 1979; Goring-Morris and Goldberg 1990).

In the Levant, loess and loess-derived deposits occur mostly along a discontinuous belt that extends from the Syrian border along the west-central part of the Transjordanian plateau to end on the western slopes of the Western Highlands in the Wadi al-Hisma region (fig. 2.3). This belt is part of a larger system, named here the "Levantine Loess Belt" (LLB), which extends from central Syria, across Jordan, and into the northern Negev and the Shephelah region in Israel.

The two largest concentrations of loess deposits along the LLB in Jordan are located in the Khan az-Zabib area and the western Hisma basin (fig. 2.3). The Khan az-Zabib Loess can be seen along the Desert Highway between Jizah and Al-Qatrana. To the west of the highway, toward Qasr Khan az-Zabib, there are various depressions and plains covered with yellow silt. Redeposited loess is found in the alluvial fills of tributary streams of Wadi ath-Thamad, Wadi ash-Shabik, and Wadi al-Mujib.

No numerical age determinations exist for loess deposits and asso-

ciated soils in Jordan. However, occurrences of Middle and Upper Paleolithic material attest to their Pleistocene origin. On the far west of the
Khan az-Zabib loess mantle, deposits on the plateaus are associated with
Middle Paleolithic and later lithics (Cordova et al. 2005). In the western
end of the Wadi al-Hisma area, near Jebel Qalkha, deep gully erosion
has exposed dozens of meters of Pleistocene deposits with sequences of
sands, silts, and paleosols (Herny 1997a) and lacustrine beds (Farhan,
Beheiry, and Abu-Safat 1989).

Eolian Sand Deposits

Quaternary eolian sands in Jordan are largely found in the southern part
of the country, mainly in Wadi Rum, Wadi al-Hisma, and Wadi Araba
(fig. 2.3). Because these are not fully open desert areas, the development
of dunes within the wadis is not as widespread as in the major sand seas of
the Arabian Peninsula and the Sahara. The typical eolian sand landforms
consist of barchan, barchanoid, echo, and climbing dunes, which can
be found stabilized or active. Stabilized dunes are evident not only because of their form and vegetation but also by their stratigraphy. In most
cases, stabilized dunes present minor pedogenic development, consisting of horizons with carbonate filaments or nodules. Dune stabilization
occurs when wind direction changes or when moisture and vegetation
stabilize sands, as has been hypothesized for the stabilized dunes in the
Ain Abu-Nukhaylah site in Wadi Rum (Henry et al. 2003).

   In the Wadi Araba, sand-dune activity is linked to sand deposited
by intense fluvial activity during the Terminal Pleistocene and Holocene
(Saqqa and Atallah 2004). Consequently, sand dune fields are found at
the end of alluvial fans descending from the mountains (fig. 2.4, top).
Dunes in the Wadi al-Hisma and Wadi Rum areas are formed under
hyperarid conditions and with enormous amounts of sand provided by
the weathering of the sandstones of the Ram Group (fig. 2.4, bottom).
Sand dunes occur in some areas of the Central Plateau, especially in areas
around the playas, where they take the form of lunettes and nebkhas.

Travertine and Tufa

These are typical carbonated deposits formed around springs. The difference between them is that tufa encompasses a larger variety of deposits
precipitated by spring water, while travertine is specifically a deposit of

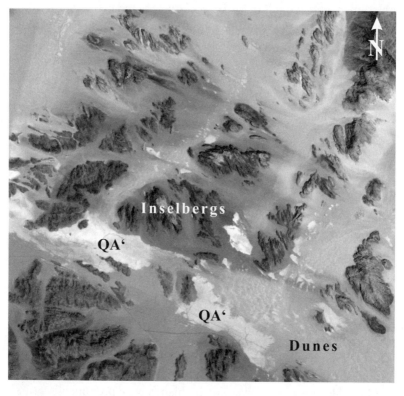

Figure 2.4 Aerial photo imagery showing (top) alluvial fans and sand dune fields in the Wadi Araba near Aqaba and (bottom) inselbergs, qaʿ basins, and sand dune fields in the Wadi Rum area. (Courtesy National Imagery and Mapping Agency © CNES/SPOT Image 1992–94.)

calcium carbonate precipitated from highly impregnated groundwater around a hot spring. Although most geoscientists use these terms synonymously, the term *travertine* is preferred here.

Quaternary travertines are found in various locations of the Central Plateau, the Western Highlands, and the Rift Valley. Travertine deposits contain fossils of plants and animals preserved in calcium carbonate. On the western margin of the Irbid Plateau, fossil-bearing travertine sequences are found in Wadi al-'Arab, Wadi Haufa, and Wadi Abu-Said (El-Radaideh 1993).

Other areas with travertine deposits of various ages are Wadi Zarqa' Ma'in (Khoury, Salameh, and Udluft 1984; Bender 1974; Banat and Obeidat 1996; Shawabekeh 1998) and the Azraq basin (Besançon, Geyer, and Sanlaville 1989), where springs have existed throughout the Quaternary, hence attracting human and animal populations. In locations along the eastern margins of the Rift Valley, travertines are often associated with lakeshore precipitation of calcium carbonate in the presence of hot water, as is the case of travertine benches in Deir Allah and the lower area of the Wadi Zarqa' Ma'in (Obeidat 1992). These benches contain large amounts of algal stromatolites, which form in environments of highly saline, shallow waters (Niemi 1997).

Calcretes

Calcretes are consolidated accumulations of calcium carbonate containing rock fragments. In soil science, calcrete is a carbonate horizon developed under dry conditions. Calcretes cap large tracts of limestone surfaces on the plateaus of the Western Highlands and the Central Plateau. In Jordan, calcrete crusts are known to the local peoples as *nārī*, which is the Arabic equivalent of the North American *caliche*. Soil scientists refer to them as K horizons, which are hardened calcic horizons often found in the lower parts of soil profiles. Despite their abundance in the Jordanian landscape, calcretes have received little attention. Therefore, little is known about their origin and ages.

Most calcretes in the Levantine highlands were presumably formed by pedogenic processes during the dry phases of the Pleistocene (Dan 1977). However, no numeric ages for them yet exist. In Jordan, six calcretes identified in the deposits of the Burma Member of the Hasa Formation were estimated as dating to 125–20 ka BP (Moumani, Alexander and Bateman 2003). These deposits mark dry periods when Lake Burma dried out. Not all calcrete deposits in Jordan are Pleistocene; some calcretes are also found associated with older deposits, as is the case of the Madaba Calc Breccia, of Oligocene age (Al-Hunjul 1995).

Geomorphology and Drainage

The division of Jordan into the physiographic regions used here is based on the originally proposed classifications of Bender 1974, later modified by Phillip G. Macumber (2001). The classification of physiographic provinces used here differs from the original in that some regions have been further subdivided into smaller units (fig. 2.5). For example, the Western Highlands and the Wadi Araba have been subdivided into smaller units based on differences in elevation, drainage patterns and lithology.

Physiographic Units

The Western Highlands present a great diversity of landscapes because of the region's large variety of rocks, elevations, and geomorphic features. It comprises a series of plateaus (e.g., the Irbid, Madaba-Dhiban, and Karak plateaus), mountains (e.g., the Ajlun Mountains), and other complex highland morphologies (e.g., the Salt-Amman Highland).

The Irbid Plateau comprises the areas between the Yarmouk River and the foothills of the Ajlun Mountains. Elevations on the plateau range from 700 meters on the northern piedmont to 400 meters on the edge of the plateau on the north and west. The main canyons dissecting this plateau are Wadi Shallalah and Wadi Aqraba (which are tributaries of the Yarmouk River) and Wadi al-ʿArab, Wadi Haufa, Wadi Hammeh, Wadi at-Tayiba, and Wadi Ziqlab (which are direct tributaries of the Jordan River).

The Ajlun Mountains occupy the area between the Irbid Plateau and the Zarqaʾ River (Wadi Zarqaʾ). The maximum elevations fluctuate between 1,000 and 1,200 meters. The Irbid Plateau and the Ajlun Mountains together are equated with the biblical historical region of Gilead, often referred to as the Gilead Mountains or Gilead Highlands. The Zarqaʾ River bounds the Ajlun Mountains to the south. In terms of annual volume of water, the Zarqaʾ River is the third largest stream in the country, after the Jordan and Yarmouk rivers.

The Salt-Amman Highland consists of rolling hills and plateaus rang-

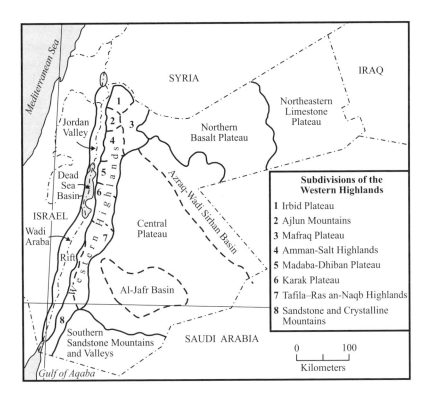

Figure 2.5 Physiographic provinces and their subdivisions (modified from Bender 1974:fig. 4 and Macumber 2001:fig. 1.1).

ing between 400 and 1,000 meters in elevation. These highlands are dissected by valleys of the Wadi Shuʿeib and Wadi as-Sir and their tributaries. This unit is now highly urbanized, as it includes the Greater Amman metropolitan area, Zarqaʾ, Suweileh, Salt, and villages now connected to larger urban areas. The biblical historical name for this region is Ammon, from which the name "Amman" derives.

The highest elevations of the Madaba-Dhiban Plateau range between 600 and 800 meters. The area is mainly agricultural and resembles the Irbid Plateau in appearance. The Madaba-Dhiban Plateau area is often referred to as the Northern Moab Plateau. The Wadi al-Mujib (the biblical Arnon River) is a deep canyon that divides the Madaba-Dhiban Plateau from the Karak Plateau. Along its nearly 800-meter depth, the Wadi al-Mujib canyon exposes a series of Cretaceous formations.

The Karak Plateau is bounded to the west by the Rift Valley, to the north and northeast by the Wadi al-Mujib, to the east by the Karak graben, and to the south and southwest by Wadi al-Hasa (the biblical Zered River). Elevations fluctuate between 800 and 1,200 meters. The topography, soils, and cultural-economic landscape are similar to those of the Madaba-Dhiban Plateau. Except for Wadi al-Karak, no major stream valleys dissect the plateau, although numerous shallow streams form a sparse drainage network on the limestone and basalt surfaces of the plateau. Very often in the historical and archaeological literature the Karak Plateau is referred to as Southern Moab or Moab proper.

The Tafila–Ras an-Naqb Highlands extend from the south margins of the Wadi al-Hasa canyon to the Ras an-Naqb escarpment. Elevations fluctuate from 1,500–1,700 meters on the west to 1,000–1,500 meters on the east. This area is part of biblical historical Edom, which also includes the Wadi Araba. Hence, very often these highlands are known as the Edom Mountains.

The Rift Valley is divided into the Jordan Valley, the Dead Sea basin, Wadi Araba, and the Aqaba Valley. The Jordan Valley and the Dead Sea basin together are also known by Jordanians as Al-Ghor. The lowest elevation in the Rift Valley is located at the shore of the Dead Sea, at an elevation of approximately −408 meters. The Jordan Valley is characterized by a wide alluvial plain and adjacent terraces formed by the Lisan Marls and other Pleistocene deposits. The shoreline of the Dead Sea presents sinuosity created by fan-deltas at the mouth of the main wadis. The Lisan Peninsula is formed by a lifted block that divides the Dead Sea basin into two sub-basins. The majority of water is contained in the northern basin, while the southern basin is in the process of desiccation. The southern basin's floor lies almost at the level of the present lake. Therefore, if the level of the Dead Sea keeps dropping, the southern basin will dry out completely and become a salt plain.

The Wadi Araba depression is here divided into two subunits: the Wadi Araba proper and the Gharandal-Aqaba depression. The Wadi Araba, the main stream draining the Wadi Araba depression, flows north toward the Dead Sea, collecting waters from the mountains to the east and from the highlands in the Negev Desert to the west. Extensive alluvial fans characterize the foot of the mountains that form the eastern edge of the valley. Some alluvial fans are inactive because they have been dislocated from the main sources of sediment by faults. The streams that descend into the depression along the alluvial fans provide a vast source of sand that is entrained by the strong winds in the valley to form sand dune fields (Saqqa and Atallah 2004). Therefore, it is not uncommon to find these sand dune fields located at the lowest end of alluvial fans (fig. 2.4, top). These sand dunes move relatively quickly, sometimes covering parts of the Safi–Aqaba highway.

The Sandstone and Crystalline Mountains extend from Ras an-Naqb to the Saudi border. This is a narrow mountain range bordering the Gharandal-Aqaba depression. The highest elevations of the mountains range between 1,300 and 1,500 meters. The base of the mountains is characterized by pediments and extensive alluvial fans on the west and east sides. The alluvial fans range in elevation from 400 to 50 meters in the west (on the Aqaba Valley side) and from 900 to 600 meters in the east (on the Wadi al-Yutm side).

The Southern Sandstone Mountains and Valleys consist mainly of isolated desert hills (inselbergs) of sandstone and valleys filled in with alluvial fan, eolian, and *qaʿ* (playa) deposits (fig. 2.4, bottom; see Osborn and Duford 1981 and Goudie et al. 2002 for details on landforms and geomorphic processes). Elevations on top of the hills fluctuate between 1,000 and 1,700 meters, and on the valley bottoms between 800 meters in Qaʿ Disa to about 1,000 meters near the Saudi border. The best area within which to appreciate the beauty and awe offered by this landscape is the Wadi Rum Nature Reserve.

The Central Plateau is subdivided into the Central Plateau proper, Al-Jafr basin, and the Azraq–Wadi Sirhan basin. The topography of the Central Plateau is highly variable, but in general it consists of rolling plains and large qaʿ depressions. Elevations fluctuate between 600 and 900 meters. On the western side, demarcating the limits between the Central Plateau and the plateau of the Western Highlands is difficult. However, the limit should be considered as generally following the divide between the waters draining into the Dead Sea basin and the interior drainages.

The Northern Basalt Plateau, also known as the Black Desert or Al-Harra, covers large tracts of land along the Syrian border. This basaltic plateau is part of the volcanic massif of Jebel Druze, whose center is located in Syria. The plateau is not completely flat; its topography varies from relatively rugged terrain, which is difficult to traverse, to relatively flat surfaces. Most of this basaltic cover has been eroded in such a way that flat-topped hills, or mesas, can be seen on its edges. The highest elevations (about 1,000 meters) are located along the northern edge of this physiographic province (see fig. 2.5).

The Northeastern Limestone Plateau, also known as Al-Hammad, is a limestone desert similar to that of the Central Plateau. Elevations in this desert fluctuate between 700 and 1,200 meters. Together with the Northern Basalt Plateau, the Northeastern Limestone Plateau forms the region known as the Eastern Badia, which is crossed by the Amman–Baghdad highway and two international pipelines. A series of depressions filled with sediments and occasionally with seasonal ponds and playas formed along the boundary between the limestone plateau and the basaltic plateau.

Surface Drainage

The hydrological basins of Jordan can be grouped under three major drainage basins: (1) the Jordan Valley–Dead Sea basin, (2) the interior basins, and (3) the drainages of the Red Sea. These major basins are further subdivided into the basins of large rivers (e.g., Yarmouk, Zarqa', Wadi al-Mujib, Wadi al-Hasa) and subsequently into the basins of smaller drainages (fig. 2.6).

The Jordan Valley–Dead Sea basin is an endorheic system; it includes the tributaries of the Jordan River and the drainages that flow directly into the Dead Sea. The largest drainages of this system in Jordan are the Yarmouk and Zarqa' rivers, both of which are perennial streams. The rest of the drainages draining the plateaus are smaller, although some of them are perennial. The largest include Wadi al-'Arab, Wadi Ziqlab, and Wadi Shu'eib. The three largest wadis emptying into the Dead Sea are Wadi al-Mujib, Wadi al-Karak, and Wadi al-Hasa. Despite the number of streams flowing into the Dead Sea, lake water levels have dropped considerably. In part this is attributable to the diversion of water and damming in the countries in the basin (Israel, the West Bank, Lebanon, and Jordan). This practice contributes not only to the drying-up of the lake but also to an increase in salinity, which is exacerbated by the chemistry of waters emanating from the surrounding springs (Abu-Jaber 1998; Capaccioni et al. 2003).

Correlation between rainfall and Dead Sea–level stands is evident through the past 150 years (Heim et al. 1997, 401, fig. 3). This means that drops and rises in water levels correspond with increased and decreased

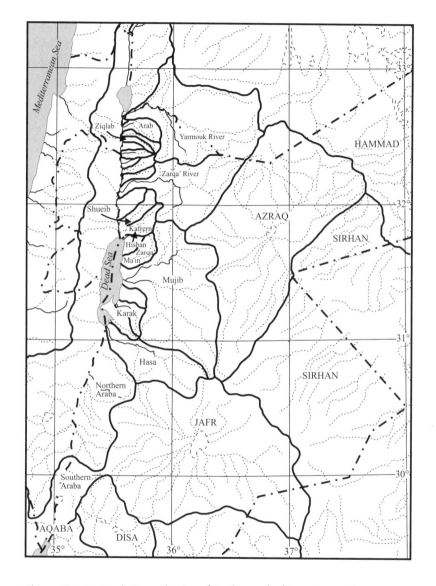

Figure 2.6 Major drainage basins of Jordan and adjacent countries.

precipitation. The synchronous behavior between rainfall and lake-level stands ended around 1975, when disturbance of water supply for irrigation and potable water use resulted in a decrease of the Dead Sea water level and an increase in brine concentration despite the increase in precipitation (Heim et al. 1997).

The economic significance of high salinity and high concentration of minerals in the waters of the Dead Sea is often underestimated. Although the lake offers no possibilities for fishing or supplying water, the Dead Sea basin represents an important source of minerals such as potash and bromide, both of which are among Jordan's export products. Other products with less economic importance include black mineral mud, which is often used for skin treatments.

The interior basins are also endorheic systems, and they often empty into dry lake beds, playas, or qaʿāt. The major qaʿ basins include Al-Azraq, Al-Jafr, Al-Jinz, Al-Hafira, Disa, and Al-Mudawwara in the Central Plateau and Qaʿ Saʿidiyin, Qaʿ Taba, and Qaʿ Defiyah in the Aqaba Valley. All the streams of these basins are predominantly ephemeral. The Azraq basin is the only one of these endorheic systems with a semipermanent supply of water, which is produced by the basaltic lava fronts in northeastern Jordan.

The only exorheic basins in Jordan are the drainages flowing into the Gulf of Aqaba (fig. 2.6). They include the drainages collected from alluvial fans and the basin of Wadi al-Yutm.

Weather and Climate

Weather Patterns

The climates of the eastern Mediterranean region and the Near East are influenced by three major circulation systems: (1) the middle- to high-latitude westerlies to the north and northwest; (2) the midlatitude subtropical high-pressure systems, which generally extend from the Atlantic across the Sahara; and (3) the monsoon climates originating in the Indian Ocean (Wigley and Farmer 1982). Because of its latitude (33°23′ N–29°10′ N), Jordan is influenced by only the first two circulation systems. However, some paleoclimatic data indicate that during some warm and humid phases of the Pleistocene and Early Holocene, the enhancement of the Indian Ocean Monsoon brought summer rains to the southern Levant (Roberts and Wright 1993; Issar 2003).

In addition to latitudinal position, continentality and topography are two significant factors controlling the distribution of weather patterns and climate types in Jordan. The amount of precipitation declines from west to east with increasing distance from the source of moisture in the Mediterranean Sea (fig. 2.7). Topography influences the distribution of precipitation by directly affecting the movement of moist air masses. The Jordan Valley and the Dead Sea basin are dry because they are in the rain shadow created by the Judean and Samarian highlands in Palestine. The descent of air into the Jordan Valley and Dead Sea basin reduces moisture but also increases air temperatures. Conversely, as air rises on the Western Highlands, moisture increases. It is for this reason that the highest amount of annual precipitation in Jordan is found on the Ajlun Mountains.

The predominant wind direction in most of the areas in the Western Highlands and the deserts of the Central Plateau is from the west and northwest, as the examples of Irbid, Safawi, and Maʿan show (fig. 2.8). Localities in the Rift Valley have a predominant northern wind created by the funneling of the north–south-trending topography of the rift depression, as is the case of Deir Allah (fig. 2.8). Thus, winds blowing from the dry and hot depressions of the Wadi Araba blow into the Red Sea, bringing heat and discomfort to the port of Aqaba.

Between October and April, low-pressure systems originate in the Atlantic and move south of the Alps and along North Africa toward the Levant and beyond by means of the westerly winds (Wigley and Farmer 1982). The relatively warm waters of the Mediterranean provide a source of moisture that feeds the storms moving into the areas inland (Roberts and Wright 1993). One of the main areas of cyclogenesis in the eastern Mediterranean is the Cyprus Low, from which cyclones move to the east and northeast and less frequently to the southeast. Consequently, the Sinai, the Negev, and southern Jordan receive much less rainfall than does the rest of the Levant. This decline of precipitation from north to south is evident along the Western Highlands. For example, regardless of their high elevation (1,000–1,300 meters), the Tafila-Shawbak Highlands receive less precipitation than do locations on the plateaus in the north at lower elevations, where elevations are usually lower than 900 meters. For example, Irbid (at 400 meters), in the north, receives 472.3 millimeters of precipitation a year, while Shawbak (at 1,000 meters), in the south, receives only 314.1 millimeters a year.

The Wadi Araba and the adjacent Negev may receive sizable amounts

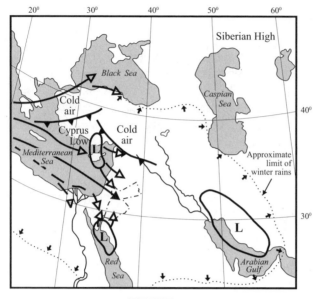

WINTER

H   High pressure
L   Low pressure
    Main paths of winter cyclones
    Less common paths of winter cyclones
    Main paths of summer cyclones
    Spring dust-storm paths
    Cold front

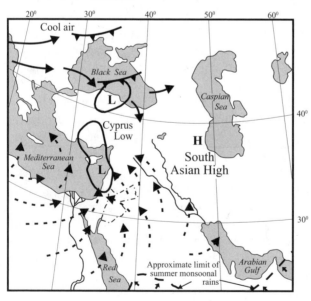

LATE SPRING–SUMMER

Figure 2.7
Jordan in the context of the main atmospheric circulation patterns of the Eastern Mediterranean and Near East.

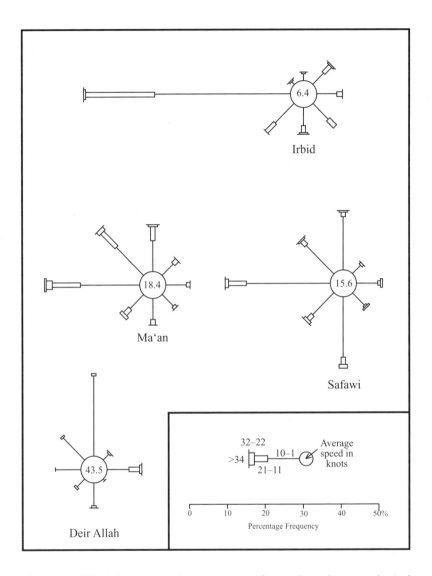

Figure 2.8 Wind direction and average speed from selected meteorological stations (see fig. 2.9 for station locations). (Data from Royal Jordanian Geographic Centre 1984.)

of water linked to the Red Sea Low (fig. 2.7). Meteorological conditions associated with events linked to the Red Sea Low do not occur every year, but detailed synoptic meteorological data allow linking enhancement of Red Sea Low events with extreme flooding events in the wadis of the Negev (Kahana et al. 2002).

Although the rainy season extends from October to April, rainfall amounts peak in January and February, when the westerlies and cold fronts intensify. The low temperatures brought by the cold fronts along with the moisture from the Mediterranean turn precipitation into snow, which in Jordan is more common at the highest elevations of the Western Highlands. During the rainy season, the number of rainy days is relatively small, since storms occur only when fronts come through (tables 2.1 and 2.2). Although cloudy days characterize the winter months, days with clear blue skies are common. Frosts are very frequent between December and March, especially in the Western Highlands and the Central Plateau.

The rest of the year (May–October) is dry. During this time, the moist westerly winds shift northward, and the belt of subsiding dry winds of the subtropical high-pressure belt covers the country. The summer is probably the quietest season weatherwise. Except for the scorching temperatures and occasional dry winds, nothing else happens during the summer days. Summer skies are commonly hazy because of the amount of dust particles in the air.

Temperatures in Jordan are mainly modified by elevation, continentality, moisture, and winds. Therefore, strong variations in mean temperature occur among the different regions of Jordan. Minimum temperatures in the highlands can be as low as −1°C, although they are normally between 3°C and 7°C, whereas in areas of Al-Ghor they can be between 8°C and 12°C. In the eastern deserts, minimum temperatures can be between −1°C and 3°C.

Maximum temperatures in the Western Highlands range between 27°C and 33°C, which is considered mild by Middle Eastern standards. Maximum temperatures in the deserts in the Central Plateau range between 33°C and 37°C, and in Al-Ghor and the Araba Valley between 35°C and 40°C. During midsummer days in Jordan, heat waves can occur, although they are not evident every year. A heat wave can raise temperatures above 40°C in Amman and as high as 50°C in some localities in the south.

The transitional seasons (spring and autumn) are characterized by relatively dry conditions interrupted by occasional rains. During these

Table 2.1 Mean monthly temperature and precipitation for selected locations in Jordan

	J	F	M	A	M	J	J	A	S	O	N	D	Yearly average
Amman													
T	9.1	10.0	12.9	17.7	21.9	25.0	26.5	26.5	25.2	21.8	15.8	10.8	18.6
P	97.5	99.9	70.8	13.1	4.9	0.1	0.0	0.0	0.4	12.4	42.2	72.9	414.2
Aqaba Port													
T	15.9	17.3	19.9	24.3	28.0	31.1	32.3	32.2	30.1	27.4	21.8	17.2	24.8
P	3.4	6.2	4.6	2.4	1.4	0.0	0.0	0.0	0.0	1.2	3.4	4.6	27.2
Deir Allah													
T	14.8	15.6	17.9	22.2	26.2	29.3	31.0	31.5	29.9	26.9	21.7	16.6	23.6
P	65.1	53.9	46.1	15.1	3.0	0.1	0.0	0.0	0.3	8.1	36.5	53.7	281.9
Irbid													
T	8.9	9.7	12.1	16.3	20.6	23.6	25.1	25.5	24.0	20.9	15.5	10.7	17.7
P	109.8	92.7	88.2	26.1	6.8	0.7	0.0	0.0	0.8	13.1	51.2	82.9	472.3
Maʿan													
T	7.6	9.0	12.1	16.9	21.1	24.1	25.6	25.8	23.9	19.7	13.6	9.2	17.4
P	7.0	7.7	7.0	3.0	2.1	0.0	0.1	0.0	0.1	4.4	4.8	6.8	43.0
Mafraq													
T	7.4	8.7	11.4	15.9	20.1	22.8	24.3	24.4	22.8	19.4	13.6	8.9	16.6
P	35.1	31.0	28.1	8.6	3.1	0.0	0.0	0.0	0.3	7.0	20.2	28.2	161.6

Sources: JMD 2005 and Weatherbase 2005.

Note: Temperature (T) in °C; precipitation in millimeters. For locations of meteorological stations, see figure 2.9.

Table 2.2 Average number of days with precipitation, highest recorded temperature, and lowest recorded temperature for selected locations

	J	F	M	A	M	J	J	A	S	O	N	D	Yearly average	Number of years
Aqaba														
N.D.W.P.	1	2	2	0	1	0	0	0	0	1	1	2	10	21
H.R.T.	34	36	36	42	43	43	45	47	42	39	35	31	47	21
L.R.T.	0	0	2	0	8	15	17	16	16	10	0	0	0	21
Irbid														
N.D.W.P.	13	14	12	6	3	1	0	0	1	5	9	13	77	10
H.R.T.	22	27	30	36	37	38	40	41	38	37	32	36	41	15
L.R.T.	−7	−3	−2	−1	4	7	11	11	11	7	1	−7	−7	15
Mafraq														
N.D.W.P.	10	9	8	3	2	0	0	0	0	2	5	10	49	10
H.R.T.	22	24	26	34	39	39	39	42	39	36	29	26	42	10
L.R.T.	−3	−6	−2	0	4	7	11	12	8	5	−2	−2	−6	8
Mahattat al-Hafif 21														
N.D.W.P.	6	5	5	2	1	1	0	1	0	1	3	4	29	14
H.R.T.	25	32	37	38	38	41	45	44	40	36	32	24	45	18
L.R.T.	−5	−3	−2	0	6	10	10	10	11	7	−1	−2	−5	18

Sources: JMD 2005 and Weatherbase 2005.

Note: Highest recorded temperature (H.R.T.) and lowest recorded temperature (L.R.T.) in °C. The abbreviation N.D.W.P. denotes average number of days with precipitation. For locations of meteorological stations, see figure 2.9. The number of years refers to the number of years during which the measurements were taken.

seasons, weather tends to be unpredictable. For example, some days in April can be as cold and rainy as January or as hot and dry as July. It is not uncommon to see intense rain and flooding in May or an early blast of polar air in October. Temperatures during the transitional season are generally mild, except during the days of seasonal dry winds. Cold mornings can be common in the Western Highlands and Central Plateau into March and April. The autumn days are often warmer than the spring days.

One of the most common weather phenomena during the transitional seasons is the regional wind known as *khamsīn*. This is a dry, dusty wind that blows from the deserts, bringing high temperatures and discomfort to most locations in Jordan and the Levant. During khamsīn days, air moisture can drop to as low as 10 percent. Usually the dusty, windy days of the khamsīn end with a cool rain that settles most of the dust in a peculiar form of mud rain. During this time of the year, temperatures are normally mild in the highlands, but the khamsīn can change this considerably, as temperatures can rise to uncomfortable levels.

Another typical seasonal wind is the *shamāl*, which may occur between June and September. It originates as a mass of polar air that sweeps over the Eurasian landmass. Therefore, this wind often blows from the north or northwest. Because the shamāl sweeps through the Syrian and Jordanian deserts, it brings clouds of dust to the Western Highlands. The wind is not too strong, and it may last for a few days. Shamāl is usually followed by a cool, light breeze and dustless air. This wind affects mainly the plateaus, as distinguished from the shamāl winds of the Wadi Araba and Aqaba region, which bring heat and discomfort to the city of Aqaba.

Climates

Jordan possesses three of the basic climate types in Köppen's classification: Mediterranean (Cs), semiarid steppe (BS), and arid (BW) (fig. 2.9). The three types have a Mediterranean precipitation regime, that is, with rainfall concentrated in the winter months.

The Mediterranean climate extends along the Western Highlands, where it can be easily delimited by the distribution of woodlands and the distribution of olive cultivation. Annual rainfall within this region is above 280 millimeters, most of which occurs between October and April. This climate type roughly corresponds to the Mediterranean floristic region (see fig. P.1, preface). The semiarid steppe climates surround the

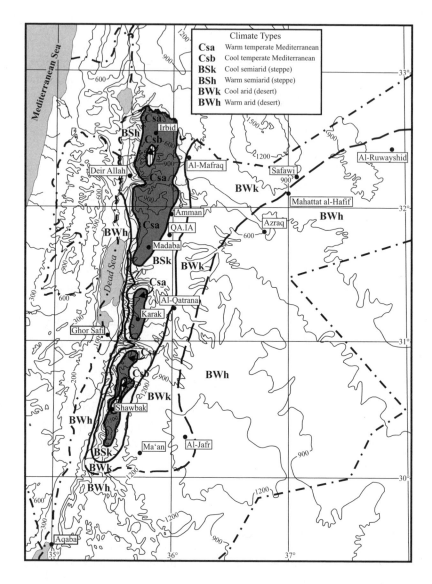

Figure 2.9 Köppen climate types and location of meteorological stations referred to in tables 2.1 and 2.2.

areas of Mediterranean climate and encompass the areas occupied by the Irano-Turanian region (fig. P.1). Temperatures vary considerably seasonally, and annual precipitation is roughly between 80 and 180 millimeters. The desert climates correspond to the arid and hyperarid zones of the rest of the country, namely, the Saharo-Arabian and Sudanian regions (fig. P.1). Temperatures vary considerably both seasonally and between day and night. Annual precipitation in the deserts of Jordan is generally below 80 millimeters.

Other climatic classifications have been applied to the territory of Jordan. The bioclimatic classification by G. A. Long (1957), later modified by Dawud al-Eisawi (1985), is often used in vegetation studies. This classification is based on the Emberger quotient (Q), which is a relation between precipitation and a combination of maximum and minimum temperatures. The advantage of this calculation is that it is designed for areas with a Mediterranean precipitation regime and areas with extreme temperatures, such as desert regions. The original classification (Long 1957) consisted of eight bioclimatic regions, but the modifications (Al-Eisawi 1985) yielded nine regions (fig. 2.10; table 2.3). Each of the major bioclimatic regions is further subdivided into a cool and one or more warm varieties, which allows differentiation of plants' temperature tolerances. The classification of these regions is useful for differentiating the locations with plants of tropical origin, such as the Sudanian elements. Likewise, the distribution of plants of higher latitudes and elevations can be easily explained through this bioclimatic subdivision.

Flora

The estimated number of plant species in Jordan varies from 2,300 to 2,500 (Hatough-Bouran et al. 1998) and includes 152 families and 700 genera (Tellawi 2001). Gymnosperms include only 3 species: *Pinus halepensis, Cupressus sempervirens,* and *Juniperus phoenica* (Tellawi 2001). The number of pteridophyta (ferns) is estimated as between 5 and 10, and the number of recorded bryophytes (fungi) and lichens so far is 150 for each (Tellawi 2001). The Royal Society for the Conservation of Nature (RSCN) estimates that in Jordan 100 species are endemic, that is to say, they are found only in Jordan (Hatough-Bouran et al. 1998).

The distribution of vegetation in Jordan is primarily influenced by the amount of annual precipitation and the temperature range and second-

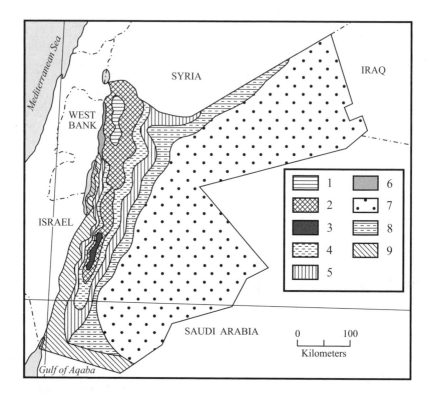

Figure 2.10 Bioclimatic regions according to Al-Eisawi 1985, with modifications. Refer to table 2.3 for characteristics of each region.

arily by lithology, soils, and topography. The interplay of these factors creates the variety of plant associations that form a complex mosaic of vegetation.

The general division of Jordan's vegetation into four floristic regions (see fig. P.1) is only an approximation of the true complexity of the country's plant geography. Naomi Feinbrun and Michael Zohary (1955) created a classification of seventeen geographic units based on dominant plant associations.

Al-Eisawi (1985, 1996) produced a classification of thirteen vegetation types (fig. 2.11), which reflects not only plant formations (e.g., forest, scrub, steppe) but also other landscape attributes (e.g., saline soils, oases, sand dunes). References to these types are made in chapters 3 and 4, where the vegetation of Jordan is discussed in detail.

Table 2.3 Bioclimatic regions of Jordan

Bioclimatic region	Emberger quotient range (Q)	Mean minimum temperature of coldest month	Mean maximum temperature of hottest month
1 Subhumid Mediterranean			
(warm variety)	70–100	3°C–5°C	< 27°C
(cool variety)	70–100	2°C–3°C	< 27°C
2 Semiarid Mediterranean			
(warm variety)	30–70	3°C–7°C	26°C–33°C
3 Semiarid Mediterranean			
(cool variety)	30–70	−1°C–3°C	26°C–33°C
4 Arid Mediterranean			
(cool variety)	10–30	1°C–3°C	28°C–39°C
5 Arid Mediterranean			
(warm variety)	10–30	3°C–7°C	28°C–39°C
6 Arid Mediterranean			
(very warm variety)	20–50	7°C–11°C	28°C–39°C
7 Saharan Mediterranean			
(cool variety)	2–15	−1°C–3°C	35°C–40°C
8 Saharan Mediterranean			
(warm variety)	2–15	3°C–7°C	35°C–40°C
9 Saharan Mediterranean			
(very warm variety)	2–30	8°C–12°C	35°C–40°C

Sources: Long 1957, modified by Al-Eisawi (1985).

Note: See figure 2.10 for geographic distribution.

Other classifications have been created for specific areas, such as the southern half of Jordan's territory (Kürschner 1986) and the Tafila-Dana area (Baierle 1993). Harald Kürschner's (1986) classification is based on bioclimatic patterns previously established by Long (1957) and on plant associations. However, this classification covers only the southern half of the country. Heinz Ulrich Baierle's (1993) classification of the Tafila-Dana area is highly detailed and based on affinities of flora (e.g., Irano-Turanian, Mediterranean). It also stresses aspects of microclimates and soil influences on plant associations, which are important ecological fac-

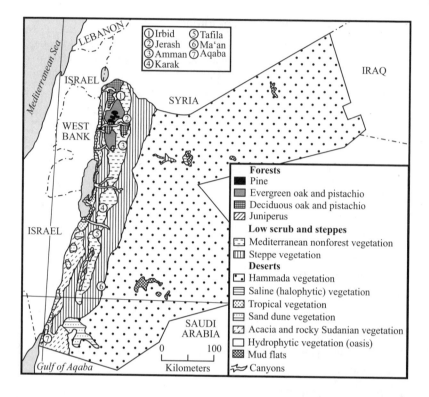

Figure 2.11 Vegetation types according to Al-Eisawi 1985, with modifications.

tors when considering the location of plants with limited distribution. Overall, these classifications reflect the high diversity of the Jordanian flora and the climatic, soil, and topographic components of the landscape.

Fauna

Mammals

The large variety in mammalian fauna represented in the archaeological records of the Holocene has shrunk considerably over time because of increased hunting (Mountfort 1965). Drawing on a diversity of historical, pictorial, and archaeological data, Alia Hatough-Bouran and Ahmad Disi (1991) were able to create a long list of large mammal species that existed in Transjordan in prehistoric and historic times. A few of the sur-

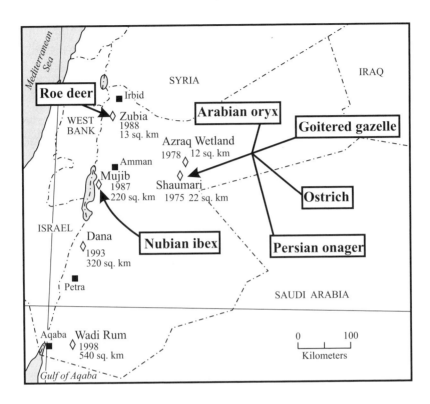

Figure 2.12 Nature reserves and reintroduced fauna. Black squares indicate cities; open diamonds indicate nature reserves.

viving species include the Nubian ibex (*Capra ibex nubiana*), wolf (*Canis lupus*), hyena (*Hyaena hyaena*), and honey badger (*Mellivora capensis*) (Hatough-Bouran et al. 1998). Other species, such as gazelle (*Gazella gazella*) and the Arabian oryx (*Oryx leucoryx*), long ago vanished from the region, although claims of sightings in several parts of Jordan suggest the contrary. Through donations from zoos abroad, most of the vanished species have been recently reintroduced and are protected in natural reserves (fig. 2.12).

The native species of large herbivores in Jordan belong to the families Bovidae, which includes gazelles, oryx, and wild goats, and Suidae, which includes wild pig (*Sus scrofa*). Carnivores include a variety of families, of which the most prominent are the Canidae (wolves, jackals, and foxes), Felidae (cats), and Hyenidae (hyenas), some of which inhabit wooded protected areas such as the Dana Nature Reserve (Amr et al. 1996). Various carnivorous species also exist in the Mustelidae family; these include martens, otters, and badgers (Zuhair n.d.).

Small mammal species are abundant, since they have had more opportunities to escape hunting. Among small mammals, squirrels, rabbits, hares, rats, and bats are the most common. Bats are a highly diverse group; they include eight families (twenty-two species in total; Zuhair n.d.), each of which has specific physical characteristics and habits, including fruit-eating bats, mouse-tailed bats, sheath-tailed bats, leaf-nosed bat, and so forth. The insectivores include two families: Erinaceidae (the hedgehog family) and Soricidae (the shrew family). Among other mammals with diverse eating habits should be mentioned the Indian crested porcupine (*Hystrix indica*), the grey hamster (*Cricetulus migratorius*), and the fat sand-rat (*Psammomys obesus*).

Birds

There are 377 species and 22 families of birds in Jordan (Amr et al. 1996). Although many are migratory birds that pass through the territory of Jordan every year, the number of native birds is high.

Ground birds include the rock and sand partridges, bustards, and ostriches (reintroduced). Flying birds include a large number of species, of which the most common are the hoopoe, bulbuls, wheatears, sunbirds, rock pigeons, turtledoves, kingfishers, rollers, and bee-eaters, among others. Birds of prey include ten species of hawks, the ornate hawk eagle (*Spizaetus ornatus*), the honey buzzard (*Pernis apivorus*), and the steppe buzzard (*Buteo buteo vulpinus*), among others.

Among the long list of migratory birds that pass through Jordan are the pallid harrier (*Circus macrourus*), the Eurasian golden oriole (*Oriolus oriolus*), the northern goshawk (*Accipiter gentilis*), the Eurasian hoopoe (*Upupa epops*), the white stork (*Ciconia ciconia*), and the European goldfinch (*Carduelis carduelis*). The Azraq wetland is one of the main stops for migratory birds in the country and thus is a bird-watching paradise and a focus area for conservation of avian fauna.

Like other groups of animals, birds have strong associations with vegetation regions. This is because vegetation forms (forests, shrublands, grasslands, etc.) represent specific habitats for specific groups of birds. In addition, cliffs and other geomorphological features represent important habitats for birds. The study by Fares Khoury (1998) on the Mediterranean wooded region of Jordan reveals the importance not only of the type of vegetation in this region but also of the type of rural activities. This study also points out that isolated woods in cliffs and inaccessible areas are refuges for endangered birds such as the lesser kestrel (*Falco namanni*) and the Syrian serin (*Serinus syriacus*). The Saharo-Arabian, Irano-Turanian, and Sudanian territories have similar exclusive habitats. Ian Andrews (1995) presents a complete list of birds of prey with illustrations and relevant information regarding their ecological contexts.

Reptiles and Amphibians

The class Reptilia in Jordan includes the Testudines (turtles), Sauria (lizards), and Ophidia (snakes). The turtles include five species in four families. Lizards and snakes are more numerous in both families and species.

Among the lizard class, zoologists distinguish the families of geckos, chameleons, agamid lizards, lacertid lizards, scinid lizards, the glass snake (*Ophisaurus apodus*) — which is not a snake but a legless lizard — and the desert monitor (*Varanus griseus*). Among the lizards is the Sinai agama, known also as blue agama or Sinai blue lizard (*Pseudotrapelus sinaitus*). This species occupies several habitats, but it is very common in the Petra area, where its blue color contrasts with the red and pink color of the rock surfaces.

The snake group includes seven families, of which the most diverse in species are the Colubridae (the largest group of snakes worldwide) and the Viperidae (vipers).

The amphibians comprise only a small group, formed by five species in five families. The salamanders, which in Jordan are represented only by the banded newt (*Triturus vittatus*), constitute one of these families. The rest of the families belong to the group of frogs and toads, which are the so-called recent amphibians.

Insects and Spiders

Among insects (Arthropoda), the butterflies (order Lepidoptera) are one of the richest orders in Jordan; they include four families and eighty-five species. The dragonflies (order Odonata) include seven families, of which the Coenagrionidae (damselflies) and the Libellulidae (dragonflies) are well represented in Jordan. The class Chilopoda (centipedes) includes eighteen species. The class Culicidae (mosquitoes) includes twenty-nine species, of which the genera *Anopheles* and *Culex* are the most abundant in Jordan.

The class Arachnida includes the ticks (order Acarina), the scorpions (order Scorpiones), and the spiders (order Aranea), among others. The ticks include the soft ticks (Argasidae) and the hard ticks (Ixodiadae), both of which inhabit camels, cows, goats, sheep, and needless to say, they harm people. The scorpions form another important order, and this is divided into three families, of which the Buthidae is the most varied, with eleven species. The Scorpionidae family has one species divided into two subspecies: *Scorpios maurus palmatus* and *S. maurus fuscus*. Spiders form a very diverse group, with a large number of species dispersed among twelve families.

Freshwater and Marine Fishes

The freshwater fauna of Jordan lacks diversity because of the scarcity of natural conditions for freshwater fish. This group of fishes includes twenty-five species in eight families. They are concentrated mainly in the Jordan River, the Azraq oases, and the numerous dam lakes of the country (Krupp and Schneider 1989). The ponds in the Azraq area contain both native and introduced species. Among the native species, the most significant is the highly endangered Azraq killifish (*Aphanius sirhani*), which is endemic to the Azraq oasis. Visitors to the Azraq Wetland Nature Reserve can see some specimens in the pond next to the visitor center. A species of catfish (*Larias lazarus*) figures also among the native fishes. The introduced fishes are two tilapids (*Tilapia zilli* and *T. aurea*) and barbell (*Barbus canis*), both of which seem to have been introduced from the lakes of the Jordan Valley (Nelson 1974).

The marine fishes in the Red Sea form a highly diverse population typical of tropical seas. There are over a thousand species in more than a hundred families in the Gulf of Aqaba, where the rugged bathymetry and reefs provide a source of food and habitat for an immense variety of marine fauna (Hulings and Wahbeh 1988). One of the main attractions in Jordan is snorkeling and diving to admire the multiple forms and colors of fish.

Nature Reserves

The idea of creating territories for the protection of endangered flora and fauna in Jordan goes back to the early 1960s, when Guy Mountfort, wildlife enthusiast and founder of the World Wildlife Fund, went on an expedition to all regions of Jordan with the purpose of gathering evidence to support the need for ecological reserves (Mountfort 1965). But not until the early 1970s did a definitive plan get under way to declare more areas under protection. The first area to obtain the official status of ecological reserve was Shaumari in 1975, followed by Azraq in 1978, Mujib in 1987, Zubia in 1988, Dana in 1993, and Rum in 1998 (fig. 2.12). These reserves are administered by the Royal Society for the Conservation of Nature (RSCN).

The Shaumari Nature Reserve straddles a territory on the transition between the Irano-Turanian steppe and the Saharo-Arabian desert. The significance of the Shaumari reserve for wildlife protection is that it is one of the natural regions where the Arabian oryx has been reintroduced. The project began when the Phoenix Zoo in Arizona provided eight individuals in 1978. Today, the number of individuals in the reserve has almost reached three hundred.

The Azraq Wetland Nature Reserve occupies twelve square kilometers of swamps and springs, which for many years have attracted birds and numerous other fauna. In decades previous to the reserve's creation, water pumping and settlement endangered the wetland, which is the largest oasis in the deserts of the Central Plateau. The restoration of the wetland was not easy, as geological water had to be pumped in. Fortunately, the project paid off, and flocks of migratory birds are now returning to the wetland. The creation of boardwalks and bird hides in the 1990s resulted in an increase in visiting birds (RSCN 2005).

The Mujib Nature Reserve covers an extensive area of canyons and slopes facing the Rift Valley. Its center is the lower reaches of Mujib canyon and the delta of Wadi al-Mujib along the Dead Sea shore. This is a nearly pristine environment of a narrow canyon in the Kurnub Sandstone. The amount of moisture represents a concentration of oasis flora. Within its 220 square kilometers and its elevation range between 900 meters and −400 meters, this region represents an area where Irano-Turanian, Sudanian, and Mediterranean vegetation come together. Its significance lies in its views, recreation, and the protection of the reintroduced Nubian ibex.

The Zubia Ecological Reserve (often known as Ajlun reserve) occupies the summits of the Ajlun Mountains. The Zubia reserve comprises a relatively dense wooded area formed by evergreen oaks, pistachio, pine, carob tree, and strawberry tree, among other rare Mediterranean plant species. This reserve provides refuge to a large number of endangered species, including roe deer, badgers, foxes, and wild boars, as well as a large number of birds. The roe deer is among the list of species now in the process of reintroduction.

The Dana Nature Reserve stretches along an altitudinal gradient comprising the southern woodlands, the escarpment, and parts of the lowlands in the Araba Valley. The Dana reserve contains a large number of rare and threatened animal species, including the sand cat, the Syrian wolf, the lesser kestrel, and the spiny tailed lizard. Like other reserves, Dana presents not only natural but also cultural attractions. Part of the project of establishing the reserve consists in preserving the Dana Village, which is one of the few places where the traditional Jordanian village can be seen.

The Wadi Rum Nature Reserve became Jordan's sixth (and the largest) nature reserve sponsored by the RSCN. Wadi Rum is among the most common ecotourist destinations in Jordan — the reason why views of this area appear in most tourist brochures and websites on Jordan. Its importance is not only ecological but also historical and archaeological. Besides the amount of rare species of desert animals and relict vegetation within its boundaries, the area has become famous through the movie *Lawrence of Arabia*, much of which was filmed using the scenic landscape now inside the reserve.

Soils

The variety of climates, topography, and rock substrate and sediment results in a plethora of soil types. Although dryland soils predominate in Jordan, soil types range from the red clay-rich soils of the Mediterranean region to the stony, highly carbonated, and highly saline soils of the desert. This geographical variation shows the importance of climate and vegetation in soil formation processes. Most soils of the Western Highlands present a xeric moisture regime, meaning that there is a deficit of water during the summer months, a typical aspect of soils in the Mediterranean region (Soil Survey Staff 1996). Soils in the desert likewise do not receive rainwater during the summer, but the desert also has a deficit of water all year long, and as a result, desert soils present an aridic moisture regime (Soil Survey Staff 1996). Tables 2.4 and 2.5 show typical soils with xeric (-xer-) and aridic (-id) moisture regimes in Jordan. Soils with xeric moisture regimes are mainly located in the area of Mediterranean climate, while those with aridic moisture regimes are found in the Irano-Turanian steppe, Sudanian, and Saharo-Arabian regions.

Table 2.4   Weighted mean values for analytical data of representative soils with xeric moisture regimes

Subgroup	Depth (cm)	Clay content (%)	CaCO$_3$ (%)	Soluble salts (ECe–mS/cm)	Number of samples tested
Typic/Entic Chromoxerert	0–25	46–48	2–22	0.27–3.0	11
(RMS)	26–60	48–61	3–23	0.31–1.60	11
	61–100	46–62	5–22	0.35–1.81	11
Vertic Xerochrept	0–25	29–57	2–21	0.36–1.07	7
(vertisol)	26–60	34–57	2–22	0.32–0.65	7
	61–100	34–59	2–23	0.32–0.59	7
Calcixerollic Xerochrept	0–25	22–48	11–50	0.36–2.02	17
(RMS)	26–60	26–53	16–55	0.32–2.36	17
	61–100	25–53	17–59	0.31–3.32	17
Lithic Xerochrept	0–25	26–43	15–34	0.59–1.17	4
(YMS)	26–60	31–50	16–34	0.57–1.46	4
	61–100	—	—	—	0
Lithic Xerorthent	0–25	28–40	23–28	0.85–0.92	3
(regosol)	26–60	—	—	—	0
	61–100	—	—	—	0
Xerochreptic Calciorthid	0–25	15–34	18–36	0.4–29.20	10
(yellow)	26–60	14–43	17–47	0.5–34.36	10
	61–100	9–43	16–62	0.5–35.70	10
Xerochreptic Camborthid	0–25	19–30	20–37	0.41–7.24	5
(yellow)	26–60	19–39	19–38	0.67–15.2	5
	61–100	21–37	16–39	1.11–17.50	5

Source: Soil Survey and Land Research Centre 1993.

Note: ECe denotes "electric conductivity of the extract." It is a measure of conductivity used to determine soil salinity, which is expressed in milliesiemens per centimeter (mS/cm). Salinity values greater than 4 mS/cm usually are damaging to crops, while salinity values less than 2 mS/cm generally permit plant growth. RMS = red Mediterranean soils; YMS = yellow Mediterranean soils.

The combination of climatic factors, geological factors, and age gives soils their main characteristics, expressed in the form of soil horizons (table 2.6). Diagnostic soil horizons not only form the basis for soil classification but also are used by geoarchaeologists and soil scientists as proxy information for reconstructing past and present processes of landscape change.

Table 2.5   Weighted mean values for analytical data of representative soils with arid moisture regimes

Subgroup	Depth (cm)	Clay content (%)	CaCO$_3$ (%)	Soluble salts (ECe–mS/cm)	Number of samples tested
Typic Calciorthid	0–25	4–45	3–49	0.73–324.32	25
(yellow soil with	26–60	3–59	3–49	0.67–124.04	25
a calcic horizon)	61–100	5–59	3–43	0.085–104.50	25
Cambic Gypsiorthid	0–25	11–43	12–40	2.92–355.84	9
(yellow soil with gypsum	26–60	10–42	9–43	16.29–294.00	9
and a cambic horizon)	61–100	10–51	9–51	16.16–115.50	9
Typic Torripsament	0–25	2–9	4–6	0.41–2.51	4
(soil developed	26–60	2–9	4–10	0.32–2.21	4
on sand dunes)	61–100	3–7	3–14	0.33–1.50	4
Typic Torrifluvent	0–25	3–31	5–55	0.48–31.60	5
(typical desert	26–60	2–32	2–69	0.43–45.56	5
alluvial fan soil)	61–100	2–39	4–55	0.41–49.00	5

Source: Soil Survey and Land Research Centre 1993.

Note: ECe denotes "electric conductivity of the extract." It is a measure of conductivity used to determine soil salinity, which is expressed in milliesiemens per centimeter (mS/cm). Salinity values greater than 4 mS/cm usually are damaging to crops, while salinity values less than 2 mS/cm generally permit plant growth.

Although the United States Department of Agriculture (USDA) tax-onomy and the Food and Agriculture Organization/United Nations Educational, Scientific and Cultural Organization (FAO/UNESCO) classifications are currently used in Jordan, this book refers to the soil types originally designated in Moorman 1959, to facilitate recognition by those readers not familiar with the jargon of modern classifications. However, table 2.7 cross-correlates the earlier soil types with more systematic classifications by providing the closest equivalents in the FAO/UNESCO classification and the USDA Soil Taxonomy. Similarly, the soil horizons used in the description of soil profiles in this book also have equivalent horizon designations (Lacelle 1986 [in the Madaba Plains Project]; Khresat 2001; Khresat, Rawajfih, and Mohamad 1998), as indicated in table 2.6.

One of the most interesting soil types in terms of prehistoric and paleoenvironmental research is the red Mediterranean soil (RMS). Although the type category "red Mediterranean soil" has been removed from most classifications because of ambiguities in geographical desig-

Table 2.6 Soil horizon designations and equivalents in other soil classifications used in Jordan

Abbrev.	Designation	Equivalents in Khresat, Rawajfih, and Mohamad 1998	Equivalents in Lacelle 1986
Ap	Plow horizon	Ap	A
A	Top organic horizon	A	A
Av	Vesicular horizon	A	A
Bw	Cambic horizon	Bw, Bwt	B
Btj	Clay-rich horizon with incipient features of illuviation	Bt, Bss	Bt
Bk	Carbonate horizon (carbonate filaments or nodules)	Bk	Cca
Btjk	Clay-rich horizon with incipient features of illuviation and carbonate nodules	Btk	
Cox	Oxidized regolith, clayey but without traces of illuviation		
K	Calcrete (caliche, *nārī*)		Cm
R	Regolith, residual deposit		

nation (Yaalon 1997), the term is still widely used to refer to a variety of red soils (Munsell color hues 5YR or redder) developed in areas with a Mediterranean type of climate. The term *terra rossa*, which is often used to refer to the red Mediterranean soils, is not used here because it may lead to confusion between primary terra rossa soil and redeposited sediments produced by erosion of those primary soils (Van Andel 1998b).

Typical colors found in these soils include dark reddish brown (5YR 3/4), reddish brown (5YR 4/4), yellowish red (5YR 4/6), and brown (7.5 YR 5/4). Their distinctive red color is the result of rubification, a typical process that occurs in soils subjected to a seasonal lack of water (xeric mois-

Table 2.7 Soil types referred to in text and equivalents in other soil classification systems

Soils designations used here[a]	Food and Agriculture Organization[b]	U.S. Department of Agriculture[c]
Red Mediterranean soils	Luvic calcisols Vertic luvisols Calcic luvisols Chromic luvisols Chromic cambisols	Xerochrepts Xerorthents Chromoxererts
Yellow Mediterranean soils	Calcic cambisols Calcic phaeozems Luvic phaeozems	Calcixerollic Xerochrepts
Yellow soils	Luvic xerosols Calcaric fluvisols Calcaric rhegosols	Camborthids Calciorthids Torriorents
Rendzinas	Rendzinas	Vary within mollisols
Regosols	Regosols	Vary within entisols
Reg soils	Lithosols Fluvisols (if in arid lands)	Torriorthents Torrifluvents Other entisols

[a] Partly based on Moorman 1959.

[b] Based on Food and Agriculture Organization 1988.

[c] Used by the Soil Survey and Land Research Centre (1993).

ture regime) that allows iron hydroxides to reach dehydration states, a process that is not easily reversible (Roquero 1993). The red Mediterranean soils are the most developed soils in Jordan. They include a series of Bt (illuvial) and Bk (calcic) horizons. Carbonates in the Bk horizons of Pleistocene soils are usually found in the form of concretions or nodules (Cordova et al. 2005).

Vertisols are often found in the plateau, particularly around Irbid. In general they are darker than the red Mediterranean soils. Vertisols often lack carbonate horizons but contain more clay than do the red Mediterranean soils.

The yellow Mediterranean and yellow soils are found in areas slightly drier than the localities of the red Mediterranean soils. The yellow soils are the typical soils of the steppe, often formed on colluvium and eolian deposits. These soils often present calcic horizons but lack argillic horizons.

Soil erosion is an important aspect studied on the plateaus of Jordan. Normally, eroded soils are associated with rain-fed agriculture and with some practices of plowing (Beaumont and Atkinson 1969).

3
Endowed Landscape
Woodland

The Mediterranean Landscapes
East of the Jordan River

The highlands east of the Jordan Rift are a wooded country in the middle of barren landscapes. Western explorers visiting this region in the nineteenth and early twentieth centuries made numerous references to the lush vegetation in contrast to the surrounding drylands (Burckhardt 1822; Wilson et al. 1881; Merrill 1883; Schumacher, Oliphant, and Le Strange 1886; Doughty 1936). Charles Doughty, who in 1876 traveled along the Hajj Road, describes the wooded lands in his *Travels in Arabia Deserta*: "Westward towards Jordan lies Gilead, a land of noble aspect in these bald countries. How fresh to the sight and sweet to every sense are those woodland limestone hills, full of the balm-smelling pines and the tree-laurel sounding with the sobbing sweetness and the amorous wings of doves! In all parts are blissful fountains; the valley heads flow down healing the eyes with veins of purest water . . . highlands of a fresh climate, where all kinds of corn may be grown to plentiful harvests without dressing or irrigation" (Doughty 1936, 55–56).

The Hajj Road runs along the steppe-desert boundary just east of this fertile land (fig. 3.1). The narrow belt of wooded land described by Doughty is the Transjordanian Mediterranean Belt (TJMB), a territory that occupies approximately 8 percent of the total area of Jordan. Higher moisture endowed this land with woods and red soils, the most fertile for rain-fed agriculture in the region. Because of its favorable natural conditions, 75 percent of the country's population lives here. The numerous archaeological sites in the area provide evidence of large concentrations of settlements in the past (Glueck 1970). Therefore, the area has been deeply transformed over the millennia, which is an aspect of the environmental history of Jordan that requires some attention.

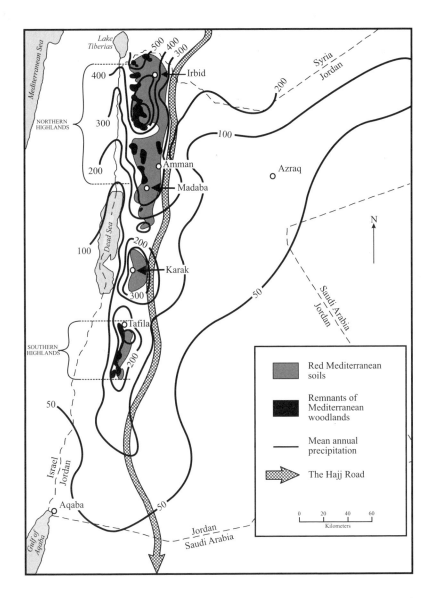

Figure 3.1 The Transjordanian Mediterranean Belt, showing mean annual precipitation (in millimeters) and distribution of woodlands and red soils.

Jordan's Woods in the Context of the
Eastern Mediterranean Vegetation

The Mediterranean vegetation type is often divided into two main zones: the Eu-Mediterranean close to the sea and the Oro-Mediterranean at higher elevations (Van Zeist and Bottema 1991). A third region is the woodland or forest-steppe region in the drier interior of the continent (Bottema and Barkoudah 1979), which is the one found in Jordan (fig. 3.2).

Annual rainfall in the Levant declines from north to south and from west to east. Differences between minimum and maximum temperatures are accentuated inland because of the decrease of moisture provided by the sea. These precipitation and temperature gradients are reflected not only in the occurrence and density of wooded areas but also in the number and geographic distribution of tree and shrub species. For example, the number and geographic distribution of pine, oak, pistachio tree, and juniper species show the different tolerances that these trees and shrubs have for dryness and extreme temperature fluctuations (fig. 3.3). The number of pine and oak species declines dramatically toward the drier and warmer areas of the southern Levant. In this distribution, *Quercus calliprinos* is the most widespread species of oak in the Levant because of its higher tolerance of drought and large variations in temperature. Like oak species, the number of pistachio species declines from north to south and from west to east. However, unlike oaks, pistachio trees tend to occupy drier areas. In some cases, pistachio trees are found in the Jordanian, Negev, and Sinai deserts, thriving in refuges where moisture is available, such as springs and streambeds. The species of juniper also decline in the same way as pistachio, but one of them, *Juniperus phoenica*, adapts to the extremely dry conditions of the highlands of the Negev and Sinai.

The *maquis* and *garrigues*, both of which are secondary communities resulting from long-term overgrazing, are typical Mediterranean plant formations that are also present in the TJMB. The maquis is a community of shrubs and small trees resulting from the degradation of forests (Dallman 1998). The maquis in the TJMB comprises a community largely dominated by evergreen oak (*Quercus calliprinos*) and Atlantic pistachio (*Pistacia atlantica*) (fig. 3.4), although variations occur depending on geological, climatic, and human factors. The garrigue vegetation type, which is also known in the Levant as *batha* (Zohary 1973), is a

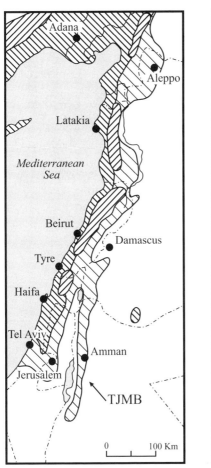
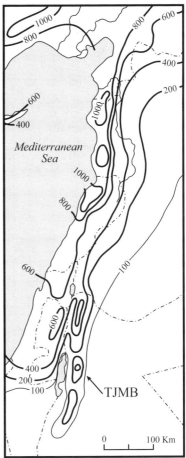

**Mediterranean Vegetation
Territories**

Oro-Mediterranean

Eu-Mediterranean

Woodland or
forest-steppe

Figure 3.2 Distribution of the main Mediterranean vegetation territories and mean annual precipitation (in millimeters) in the Levant.

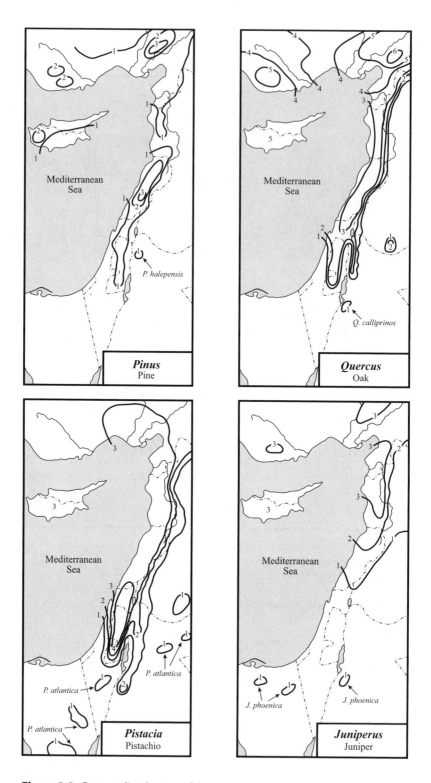

Figure 3.3 Current distribution of pine, oak, pistachio, and juniper by number of species in the Levant.

Figure 3.4 Maquis vegetation around Ajlun Castle.

community of perennial low scrub resulting from the degradation of the maquis by means of intense overgrazing and frequent fires (Naveh and Dan 1973). Garrigues are in general composed of antipastoral plants, or plants avoided by livestock. In the TJMB, thorny burnet (*Sarcopoterium spinosum*) and thorny broom (*Calycotome villosa*) form the most common association of garrigue vegetation (fig. 3.5). Other associations of garrigues include plants of the mint family (Labiatae), among which *Ballota undulata, Salvia dominica, Astragalus bethlemiticus*, and *Marrubium libanoticum* are the most abundant (Feinbrun and Zohary 1955).

The Two Main Wooded Areas

The natural Mediterranean arboreal communities of the TJMB are concentrated in two regions, referred to as the northern and southern highlands (fig. 3.1). The northern highlands include slopes facing east along the edge of the Irbid Plateau, the Ajlun Mountains, and the area south of the Zarqa' River to the towns of Salt and Sweileh. The southern highlands include the escarpment and western margin of the plateau between Tafila and Wadi Musa. In the historical and archaeological litera-

Figure 3.5 Garrigue vegetation dominated by *Sarcopoterium spinosum* (prickly shrubby burnet) near Salt.

ture, the northern highlands correspond to the Mountains of Gilead and the southern highlands to the Mountains of Edom. Arboreal communities have remained in these two regions primarily because of the relatively high amount of precipitation, usually above 300 millimeters a year, which is enough to maintain woodlands.

Differences in topography, lithology, soil, land use, and plant associations between the northern and southern highlands of Jordan are illustrated along an altitudinal transect (fig. 3.6). The escarpment on the southern highlands is clearly steeper, which may not have a direct impact on the distribution of vegetation but is likely to result in different forms of land use because the land is too steep for agricultural activities. The gentler topography of the northern highlands allows higher population densities, which implies a faster reduction of the wooded communities (Chapman 1947). Rapid population growth in the northern highlands transformed large tracts of former wooded areas into olive tree plantations or into grazing grounds.

Lithology and soils affect the distribution of certain plant species. Well known is the wide distribution of *Juniperus phoenica* in the southern highlands and its absence in the northern highlands (Al-Eisawi 1996).

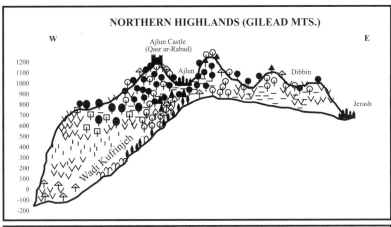

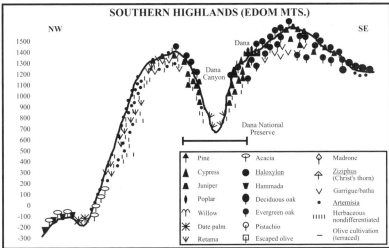

Figure 3.6 East–west transects across the northern and southern highlands.

This species prefers sandy soils, especially those produced by the Nubian Sandstone formations, which are exposed only in the south (Kürschner 1986; Baierle 1993; Al-Eisawi 1996). In contrast, madrone (*Arbutus andrachne*) is absent in the southern highlands. This probably has to do with a similar preference, in this case for soils developed on limestone. However, climatic differences are also evident between the two regions. Precipitation is lower in the southern highlands, despite the high elevations there. But the most contrasting characteristic is winter temperatures, which account for bioclimatic differences. The northern highlands comprise the cool variety of the subhumid Mediterranean bioclimatic region, where January minimum temperatures fluctuate between 3°C and 7°C, whereas the southern highlands belong to the cool variety of the semiarid Mediterranean bioclimatic region, where January temperatures fluctuate between −1°C and 7°C (see table 2.3). Perhaps the strong presence of *Artemisia herba-alba* in the southern highlands is accounted for by the relatively drier and cooler climate.

The area between the two wooded areas in the northern and southern highlands is referred to by Dawud al-Eisawi (1985, 1996) as the Mediterranean nonforest vegetation (see fig. 2.11). This area includes the entire territory of the Madaba, Dhiban, and Karak plateaus, which are covered with cultivated fields and various communities of scrub and herbaceous vegetation. Associations of *Salvia dominica–Ballota undulata* and *Astragalus bethlemiticus–Marrubium libanoticum* dominate the garrigue vegetation of this part of Jordan (Feinbrun and Zohary 1955). Although the annual precipitation in most areas is between 300 and 400 millimeters — enough to maintain woodlands of oak and pistachio — the region lacks native tree growth.

Several isolated areas of scattered woods exist in the nonforest Mediterranean vegetation area. One of them is in the area of Naʿur, near the road from the capital to the Jordan Valley (Bardtke 1956; Long 1957; Crawford 1986). The other one is the Ataruz wooded grassland, discussed below.

Common Trees and Shrubs

When the Mediterranean climatic rhythm (summer drought) began approximately 3.3 million years ago, a selection of plants appeared, thus causing the individualization of the modern Mediterranean elements (Pignatti 1978; Suc 1984). This selection of plants from diverse origins (west, east, tropical, and middle latitude) succeeded in overcoming the environmental challenges imposed by the dry summers and wet, cold winters characteristic of the Mediterranean climate (Mitrakos 1980).

The challenge to adapt to a Mediterranean climate implies that trees and shrubs have to cope with high evapotranspiration rates during the long, dry summer period, when temperatures are high and days are long. Water is available in the winter, but temperatures are low and daylight is shorter, limiting plant growth. For this reason, vegetation in the regions with Mediterranean climate represents a unique set among the world's natural regions (or biomes).

The environmental challenge described above is even greater in marginal areas of the Mediterranean biome, where proximity to the deserts implies harsher living conditions, with decreased amounts of precipitation and more extreme temperatures, as is the case of the TJMB. Despite difficult growing conditions, many Mediterranean plants managed to adapt to the climatic marginality of the TJMB, hence creating a diversity of trees and shrubs.

Oak

Of the thirty species of oak in the eastern Mediterranean region, only two are among the native flora of Jordan: the evergreen Kermes oak (*Quercus calliprinos*) and the deciduous Tabor oak (*Q. ithaburensis*) (Zohary 1973). Each of these species includes several varieties, which in some cases have been identified as different species (Zohary and Feinbrun-Dothan 1966; Al-Eisawi 1996).

Quercus calliprinos is more tolerant of dryness, cold, and heat, and its soil requirements are less demanding than those of *Q. ithaburensis* (Feinbrun and Zohary 1955; Long 1957; Al-Eisawi 1985, 1996). Pollen diagrams from the lakes of the Rift Valley (Baruch 1990; Baruch and Bottema 1999) and the Golan Heights (Schwab et al. 2004) show an inverse development of the curves of deciduous and evergreen oaks (see fig. 5.2). The most notorious case of this inversion occurs after the Roman-Byzantine period; following the decline of olive pollen, deciduous oaks never recovered but evergreen oaks increased. This could mean also that evergreen oaks are more capable of colonizing disturbed areas.

Quercus calliprinos is by far the most widespread of all the oak species in the Levant (Zohary 1973). It occupies wider elevation ranges than other oak species. In Jordan, *Q. calliprinos* forms associations with *Pistacia palaestina* and *P. atlantica*, although it is often found in isolated areas where farming expands at the expense of former woodlands. In the southern highlands, *Q. calliprinos* marks the southernmost limit of oaks in Jordan and the eastern Mediterranean region (Zohary 1973).

Quercus ithaburensis, also known as *Q. infectoria* and *Q. aegilops*, is found only on the northern highlands, where it grows in association with *Pistacia atlantica, P. palaestina, Crataegus azarolus, Rhamnus palaestinus*, and a variety of other shrubs. No data suggest that *Q. ithaburensis* ever existed in the southern highlands, where colder and drier conditions may limit its growth.

Even though today's oaks are among the most important arboreal constituents of the Levant's woodlands, oak tree cover has been considerably reduced throughout the entire history of agriculture in the region. The pollen diagrams from the Sea of Galilee (Baruch 1990), Hula Lake (Baruch and Bottema 1999), and the Ghab Valley (Van Zeist and Bottema 1991; Yasuda, Kitagawa, and Nakagawa 2000) show this trend. Therefore, it is possible to think that the stands of oaks we see today in the TJMB and elsewhere in the Levant are just small remnants of the once extensive forest and woodlands of oak.

Pistachio

The genus *Pistacia*, known as pistachio or terebinth, has an interesting historical and geographical background. *Pistacia* was among the important constituents of the Mediterranean flora during the Pliocene (Pignatti 1978; Suc 1984). Today, the modern distribution of *Pistacia* is subtropical, occurring predominantly in the Northern Hemisphere. With the exception of two North American species (*P. mexicana* and *P. texana*), the genus *Pistacia* is mostly concentrated in Europe, Asia, and Africa (Zohary 1952). Nine of the twelve species are concentrated within latitudes of 20°–40° N. Outside this range, one species is found as far north as southern Crimea, and one as far south as Tanzania (Zohary 1952).

Today, only seven of the twelve species of *Pistacia* are found in the Mediterranean region, where they are an important component of the maquis vegetation. Three species of *Pistacia* grow wild in Jordan: *P. atlantica, P. palaestina,* and *P. khinkjuk* (Feinbrun and Zohary 1955; Al-Eisawi 1996). Other species, such as *P. vera* and *P. lentiscus,* are introduced and cultivated for their edible nuts.

Pistacia atlantica is the most widespread species of pistachio in Jordan. In the TJMB, *P. atlantica* forms associations with *Quercus calliprinos* in both the northern and southern highlands. In the drier parts of the TJMB, *P. atlantica* forms associations with *Crataegus azarolus* and *Rhamnus* spp. Very often pistachio trees are found as the main constituent of the forest-steppe region or forming isolated patches in savanna-like vegetation as relict stands of former woodlands (fig. 3.7).

Pollen records from the bottom of the Mediterranean Sea show an increase in the overall amount of *Pistacia* pollen between 9000 and 6000

Figure 3.7 Scattered relict stands of *Pistacia atlantica* (Atlantic pistachio) near Ataruz.

[14]C years BP, indicating rising temperatures and winter precipitation (Rossignol-Strick 1999). After 6000 BP, the amount of *Pistacia* pollen dropped in the marine records but increased in several small basin pollen diagrams, such as southwestern Turkey (Pons 1981) and northeastern Greece (Bottema 1991). This late expansion of *Pistacia* is assumed to be the result of widespread cultivation of *P. lentiscus* by Greek farmers (Huntley and Birks 1983). Pollen records from Beyşehir Gölü, Turkey, show *Pistacia* as one of the species increasing after the expansion of early farming at the beginning of the Beyşehir phase (ca. 4000 BP) (Bottema and Woldring 1990). In a different temporal and geographical context, pollen records from southwestern Crimea show *Pistacia* as one of the main trees expanding after the abandonment of Greek farms after the second century BC (Cordova and Lehman 2003). Therefore, the expansion of pistachio trees may also be the result of human disturbance.

Pollen diagrams from Hula Lake (Baruch and Bottema 1999) and Lake Kinneret (Baruch 1990) in northern Israel, which have been used as proxies for the paleovegetation of the TJMB, do not show considerable amounts of pistachio pollen. Indeed, the levels of *Pistacia* pollen in modern samples are always low, reflecting poor dispersion and preservation (Huntley and Birks 1983; Rossignol-Strick 1998). In addition to their rapid resilience after they have been cleared, pistachio trees are also resistant to dryness. This explains why pistachio trees are among the wooded species thriving in some enclaves in the deserts of the southern Levant (Danin 1983; Al-Eisawi 1996). Pistachio trees can be found in isolated stands along wadis in drier areas outside the Transjordanian Mediterranean Belt (fig. 3.8).

Olive

Like *Pistacia*, the genus *Olea* has been an important element of the Mediterranean flora since the Tertiary period (Raven 1971; Pignatti 1978; Suc 1984). Both species then became domesticated and widely distributed in plantations around the Mediterranean region. Today, the distribution of olive cultivation defines the boundaries of the Mediterranean climate (King 1997; Dallman 1998). The widespread cultivation of olive trees since the early stages of agriculture has overshadowed the role of *Olea* as a wild element in the Mediterranean flora. Despite their strong presence as cultivated species, wild olives are still an important constituent of the maquis vegetation around the Mediterranean (Zohary 1973).

Wild olive tree stands have been reported in several locations in Jordan, including the Zarqa' River valley, Wadi Kufrinjeh, Wadi Rajib, and other valleys draining the Ajlun Mountains (Steurnagel 1925; Hatough-Bouran et al. 1998; Neef 1990). Their existence in these refuges suggests that olive trees may have been part of the original Mediterranean woodland, just like oak and pistachio. However, some authors agree that the so-called wild varieties are escapees from cultivation, namely, trees that have reverted to the wild as they were abandoned (Neef 1990). Despite this disagreement, the wild olive (*Olea europaea* var. *oleaster*) is regarded as the ancestor of the domesticated olive (*Olea europaea* var. *europaea*) (Zohary and Spiegel-Roy 1975; Neef 1990). Differences between the two varieties exist, as the wild variety presents smaller fruit size and lower oil content (Zohary and Spiegel-Roy 1975). Further differentiation poses a problem because stone sizes overlap and other distinctive characters contrasting the two are not clear (Zohary and Spiegel-Roy 1975; Neef 1990), in addition to the fact that wild olive is fully interfertile with cultivated varieties, leading to cross-breeding.

The problem of differentiating wild from cultivated varieties becomes even more problematic when interpreting olive remains (wild or cultivated) recovered from early farming archaeological sites (Liphschitz et al. 1991). One example of this problem is evident in olive stones and wood findings from the Chalcolithic site at Tuleilat el-Ghassul, located in the Jordan Valley at 290 meters below sea level. With precipitation under

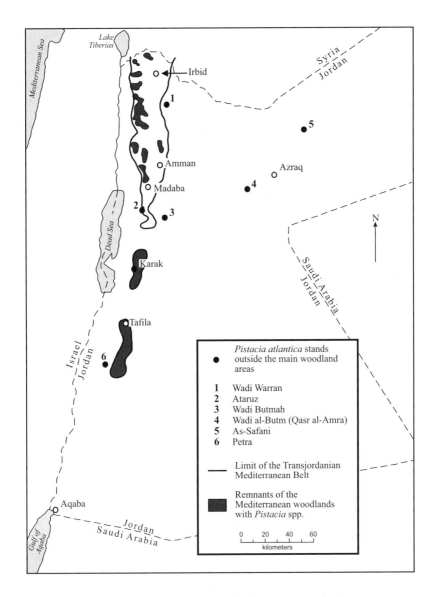

Figure 3.8 Relict stands of *Pistacia* outside the main wooded areas.

100 millimeters a year, this site falls below the distribution of wild olive trees, which ranges between 300 and 800 meters of elevation and between 300 and 500 millimeters of annual precipitation (Neef 1990). This geographic discrepancy leads some researchers (Zohary and Spiegel-Roy 1975) to believe that the stones and wood recovered here are more likely to have occurred through cultivation, raising the question of whether irrigation was practiced in the area. Floodwaters from the nearby wadis originating in the highlands may have been used for irrigation. However, no convincing evidence to support irrigation has been found around the site.

Another clue suggesting the rarity of olive trees in the wild in preagricultural times lies in some of the Levantine pollen assemblages dated to the Early Holocene. The pollen diagram from the Ghab Valley in northwestern Syria shows that after 7750 ^{14}C years BP, *Olea* pollen gradually began to outnumber the pollen of other trees, presumably indicating the expansion of olive cultivation (Yasuda, Kitagawa, and Nakagawa 2000). This date coincides with the increase of farming communities during the Pottery Neolithic period. Unfortunately, as in the case of archaeobotanical remains, the domesticated and wild varieties cannot be differentiated through pollen analysis.

Despite difficulties in interpreting early domestication, it is evident that olive trees, whether domesticated or wild, are an important component of the Mediterranean landscapes of Jordan. As in the rest of the Mediterranean borderlands, olive plantations mark the limit of the Mediterranean climate in Jordan; scattered olive plantations exist in the surrounding drylands only under irrigation.

Aleppo Pine

Despite the large number of species of pine in the Mediterranean region, Aleppo pine (*Pinus halepensis*) is the only native species of pine in Jordan (fig. 3.3). Although Aleppo pine has been given other names (e.g., *Pinus hierosolimitana* Duh., or Jerusalem pine), *Pinus halepensis* Mill. has been accepted by most botanists (Zohary and Feinbrun-Dothan 1966). Named after the city of Aleppo in northwestern Syria, this species is rather rare in Syria and in the entire eastern Mediterranean region. Aleppo pine is actually a pan-Mediterranean species with a stronger presence in the western Mediterranean, especially in Spain and North Africa (Liphschitz and Biger 2001). The species designation *halepensis* originated in 1876, when this pine was described in one of the public gardens in the city of Aleppo where pines were planted (Liphschitz and Biger 2001).

In Jordan, wild stands of Aleppo pine are confined to the Ajlun Mountains (fig. 3.6), occupying areas above 700 meters in elevation with annual precipitation above 550 millimeters and calcareous soils (Al-Eisawi 1996; Hatough-Bouran et al. 1998). Elsewhere in Jordan, the occurrence of pine is solely attributable to recent reforestation (Al-Eisawi 1996).

The ecological role of Aleppo pine in the Ajlun Mountains is questionable, and some researchers have suggested that its presence is the result of frequent fires (Feinbrun and Zohary 1955; Long 1957; Zohary 1973; Al-Eisawi 1996). Studies on Aleppo pine propagation in Palestine have linked it to the history of fire in the previous century (Kutiel and Naveh 1987; Kutiel 1994). Archaeological and historical-geographical data indicate that before the twentieth century, Aleppo pine was rare in Palestine (Liphschitz and Biger 2001). References to Aleppo pines by travelers visiting Palestine and Transjordan in the nineteenth century are rare. Although Doughty (1936) made reference to pine stands in the Western Highlands, extensive surveys of vegetation in Transjordan (Schumacher, Oliphant, and Le Strange 1886; Steurnagel 1925) acknowledge the poor representation of pines.

Palynological and paleobotanical evidence show minimal presence of pine in the region in prehistoric and historic times (Liphschitz and Biger 2001). However, some pollen records elsewhere in the Near East suggest that the areal extent of pine expanded after periods of land clearance. In a pollen core from the sediments of Lake Söğüt in Turkey, a considerable increase in pine pollen after a phase of deforestation between 2,900 and 1,500 years ago has been identified (Van Zeist, Woldring, and Stapert 1975). Similarly, the pollen core from the Ghab Valley in northwestern Syria shows an increase of pine as other trees (e.g., cedar and deciduous oaks) decline, presumably as a consequence of early agricultural clearance during the Pre-Pottery Neolithic (Yasuda, Kitagawa, and Nakagawa 2000). Farther south, in the pollen diagrams of Lake Hula (Baruch and Bottema 1999) and Lake Kinneret (Baruch 1990), pine pollen values are always low, even in the context of agricultural intensification.

Perhaps the rarity of pines in Palestine and Transjordan lies in their marginality with respect to the distribution of Mediterranean pines. According to Michael Zohary (1973), the stands of Aleppo pine in both Palestine (the Hebron area) and Transjordan (the Ajlun Mountains) mark the southeastern limit of the species' distribution in the Mediterranean realm (fig. 3.3) and the southernmost limit of the genus *Pinus* in the Mediterranean region. This marginal location suggests that Aleppo pine in this part of the Mediterranean may occur either as a relict species or as a recent invader. The possibility that Aleppo pine is a relict implies that this species may have been abundant in the region in past geologic periods and that it was left in its present refuges in the Ajlun Mountains as moister and colder climates receded. However, no fossil evidence exists to support this assumption. The possibility that Aleppo pine is a recent invader can be explained on the basis that this species has occupied the niches of other trees removed by humans, as Zohary (1973) has postulated.

Phoenician Juniper

The only species of juniper reported for the native vegetation of Jordan is *Juniperus phoenica* (Phoenician juniper), existing only in the southern highlands of the TJMB (figs. 3.3 and 3.6). In the eastern Mediterranean region, this species of juniper has an odd distribution, for it can be found in extremely dry areas in the Sinai and in moister areas of the mountains in Cyprus (Zohary 1973). Its preference for soils is another inexplicable oddity; its presence in southern Jordan relates to the sandy soils of the Nubian Sandstone group (Kürschner 1986; Al-Eisawi 1996), while in the Sinai, Cyprus, and other areas of the Mediterranean, it grows on limestone-derived soils or any other soil type. Why, then, is it not present on the limestone slopes of the northern Jordanian highlands? One possible explanation is that microclimatic factors in addition to the sandy soil preference are involved in its distribution. The tolerance of *J. phoenica* for dryness and cold is well known. In the southern highlands, Phoenician junipers are found at elevations as high as 1,500 meters with slightly below 300 millimeters of rain a year and temperatures that frequently drop below 0°C.

Hawthorn

Only two species of hawthorn are common in Jordan: *Crataegus azarolus* and *C. aronia*. Hawthorns are typical Mediterranean shrubs with strong resistance to dryness and human-induced disturbance, making them most suitable to the driest part of the Mediterranean region, namely, along its limits with the Irano-Turanian steppe (Zohary 1973).

When hawthorns occur in wooded areas, they are usually associated with *Quercus calliprinos, Q. ithaburensis*, and *Pistacia* spp. When outside the wooded areas, they are typically associated with garrigue vegetation (*Sarcopoterium spinosum* and *Calycotome villosa*). Hawthorns are reliable indicators of Mediterranean borderlands of former steppe-forest (Zohary 1973), hence suggesting their resistance not only to dryness but also to overgrazing. In particular, *C. azarolus* is the most widespread species of hawthorn in the nonforest Mediterranean vegetation zone. Both species of hawthorn produce edible seeds, but only *C. aronia*, the tastier of the two, is sometimes cultivated (Zohary 1973).

Madrone

The species of madrone in Jordan is *Arbutus andrachne*, a tall shrub found in only a few areas of the northern highlands. Madrones are relatively modest in their ecological demands, preferring thin calcareous soils (Zohary 1973). Like that of olive trees, the distribution of madrones is a good marker of the geographical extent of the Mediterranean climate (Rikli 1943; King 1997; Dallman 1998). The distribution of Mediterranean madrones is divided into two main domains: *A. andrachne* in the eastern Mediterranean and *A. unedo* in the western Mediterranean, with the latter extending slightly farther into the eastern Mediterranean, as far as western Turkey (Rikli 1943; Zohary 1973).

Madrones in the Ajlun Mountains form associations with *Quercus calliprinos, Pinus halepensis*, and *Cupressus sempervirens*, but very often they occur in pure stands (Zohary 1973). The madrone stands in the upper reaches of Wadi Kufrinjeh seem to be associated with *Pinus halepensis, Cupressus sempervirens* var. *horizontalis*, and *Cistus villosus* (Feinbrun and Zohary 1955).

Oleander

Nerium oleander is an evergreen shrub typically found along permanent stream courses and around springs. Although oleander is a typical circum-Mediterranean shrub, oleanders are also found along wadi courses in the Irano-Turanian steppe. However, the extension of this species into the deserts, in particular the Saharo-Arabian region, is apparently restricted by the depth and variability of the water table. Oleander leaves are highly poisonous and avoided by livestock, which is why oleander is more abundant than other typical riparian plants that are often eaten by goats.

Cypress

The only species of cypress in the woodlands of Jordan is *Cupressus sempervirens* var. *horizontalis*, known as the Italian cypress. Although this species has a wide distribution in the Mediterranean region, its occurrence in Palestine and Transjordan is rare. They are scattered in the northern highlands, with Wadi Kufrinjeh being one of the best-known communities, where *C. sempervirens* var. *horizontalis* occurs in association with *Pinus halepensis* and *Arbutus andrachne* (Zohary 1973). Its presence in the southern highlands is restricted to one locality near Tafila, where cypress trees more than a thousand years old (Al-Eisawi 1996) indicate a relict community. The locality is known as Khunag al-Arz (*arz* is Arabic for cedar), because the trees were thought to be cedars by elderly local people (Al-Eisawi 1996). Its rarity in the woodlands and its presence in isolated spots suggest that cypress in Transjordan is receding from a larger geographical distribution. Unfortunately, the history of cypress in the Levant is difficult to assess because its pollen grain is usually determined only to the family level (Cupressaceae), which also includes juniper.

Almond

Three species of almond are common in Jordan: *Amygdalus communis*, *A. korschinskyi*, and *A. arabica* (Al Eisawi 1996). Like *Pistacia*, *Amygdalus* is a genus shared by the Mediterranean and Irano-Turanian provinces (Zohary 1973). Therefore, almond trees are common in wadi bottoms in the Irano-Turanian steppe and in a few cases in the Saharo-Arabian province.

Although *Amygdalus communis* is one of the most widespread species of almond in the Middle East, its distribution in the woodlands of Jordan is minimal (Al-Eisawi 1996). *Amygdalus communis* L. var. *amara* is the wild variety of almond, while *Amygdalus communis* L. var. *dulcis* is the domesticated one (Zohary 1973). While the wild variety is rare, the cultivated variety is found almost everywhere in the TJMB, usually in family gardens and around cultivated fields. The wild almond tree grows in association with oak and pistachio. In the southern highlands, stands of *Amygdalus korschinskyi* are often found in association with *Juniperus phoenica*. In Jordan, *A. arabica* is rare and occurs mostly in wadi bottoms in the desert (Al-Eisawi 1996).

Carob Tree

Ceratonia siliqua is an evergreen tree typical of the eastern Mediterranean region, but its occurrence east of the Jordan River is rare (Zohary 1973). Even in Palestine it is relatively rare in the wild and rare in the archaeobotanical record. Charred wood and seeds of this tree are rare in archaeological deposits and more common in recent centuries, which is an indication that carob trees are newcomer species that invaded denuded areas where natural vegetation had been destroyed by humans (Liphschitz 1987).

In Jordan, carob trees are found in a few areas between the Yarmouk River and the Ajlun Mountains (Steuernagel 1925; Long 1957), where it is found in association with *Olea europaea* and *Quercus ithaburensis* (Long 1957). Carob trees were formerly believed to be absent from the southern highlands, but recently they have been reported in the Dana Nature Reserve.

Were There Any Forests in Jordan?

Archaeobotanical remains recovered from numerous early agricultural sites suggest that trees existed in the nonforest Mediterranean vegetation zone in prehistoric times. However, these remains only provide a list of plants that were used by the prehistoric settlers at a particular location; little is known about the structure and coverage of the arboreal communities around the sites.

Fossil pollen records provide proxy information on coverage and structure of forests, as well as the sequence of change. But pollen records for the TJMB are either not available or are fragmentary. With the lack of continuous proxy data on paleovegetation, other resources indirectly aid the reconstruction of ancient vegetation. One of these resources is the pollen records of neighboring regions, which provide information on paleoenvironmental change in areas with physical characteristics similar to the landscapes of Jordan. The other resource lies in the modern landscape, which contains clues to past vegetation change. Thus, the distribution of modern vegetation communities with respect to climate, soil, rock, and land use provides tools for reconstructing the landscape. Thus, before answering the question "Were there forests in Jordan?" it will be necessary to first answer the following questions: (1) Are there any forests in Jordan today? and (2) Were there ever appropriate conditions that could have sustained forests in Jordan?

Are There Any Forests Today in Jordan?

To seek an answer to this question, we must first look at those areas of Jordan that we call "forests." The term *forest* can be sometimes vague, for we use it as a general term to designate wooded vegetation communities that are not necessarily forests. To clarify this point, the concept of potential natural vegetation must be distinguished from actual vegetation.

Potential natural vegetation is the type of vegetation that would grow in an area if there were no human disturbance (Küchler 1988). The best example of this concept is shown by the classifications of modern vegetation for the Levant (Zohary 1973; Van Zeist and Bottema 1991) and for Jordan in particular (Al-Eisawi 1996; Kürschner 1986). The potential natural vegetation represented in these maps is just an approximation of what type of vegetation should exist under certain climatic, topographic, and soil conditions. In contrast, the actual vegetation is a mosaic of plant communities that we see on the ground or on an aerial photograph. Actual vegetation can be "natural," "seminatural," or "cultural," depending on the degree of human transformation.

One example of the difference between potential natural vegetation and the actual vegetation (seminatural and cultural) is illustrated in an example from the northwestern edge of the Irbid Plateau (fig. 3.9). Al-Eisawi (1985, 1996) classifies this area as "deciduous oak forest." This designation implies that according to elevation and the general climatic and soil conditions, deciduous oak forests would occupy this area. But interpretation of vegetation and land use on an aerial photograph shows a different picture. The actual vegetation of the Wadi Haufa area looks nothing like a forest but is instead a mosaic of seminatural vegetation (e.g., woodland of deciduous oak) intermingled with cultural vegetation (e.g., olive groves and cultivated fields) (fig. 3.9-B). The patches with woods in this area have tree canopy coverage that is generally less than 40 percent (fig. 3.9-C). The minimum limit for a forest in most classifications of vegetation (according to Zonneveld 1988) is 80 percent. There-

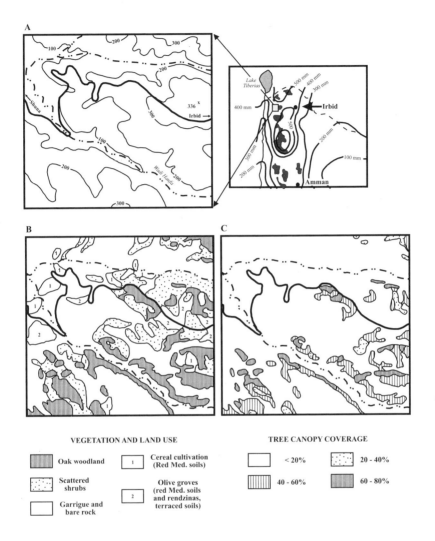

Figure 3.9 (A) Location and precipitation gradients of the Wadi Haufa area on the northwestern edge of the Irbid Plateau; (B) actual seminatural and cultural vegetation and land use; and (C) tree canopy coverage.

fore, the wooded areas in the area shown in fig. 3.9 are not dense enough to be classified as forest.

Most of the presently wooded areas of the TJMB are patches of open woodlands of the maquis type similar to the one presented in the example above. Even in the relatively denser wooded areas of the Ajlun Mountains and in protected areas such as the Dana reserve, the tree canopy coverage is usually below 80 percent. In most vegetation mapping classifications, if tree canopy cover is less than 80 percent but above 40 percent, the community is termed "woodland," and if tree cover is below 40 percent, the community is called "wooded grassland" or "savanna" (Zonneveld 1988). Therefore, in a strict sense there are no forests in Jordan today. Instead, the actual vegetation of the so-called forested areas is in the category of degraded woodlands (of the maquis type). However, the possibility that moister periods created conditions for the development of forests in preagricultural times is an issue that has to be examined in the biogeographical context of paleobotanical data.

Were There Ever Appropriate Conditions for Forests in Jordan?

Pollen data in other areas of the Levant, such as the Galilee (Lake Kinneret and the Hula basin), suggest that during some times in the Terminal Pleistocene and Early Holocene, areas around the pollen core sites were densely forested (Van Zeist and Bottema 1991; Baruch and Bottema 1999). This is possibly true for areas at higher elevations, where higher precipitation amounts supported a forest cover, especially under conditions of reduced human impact. This could also be the case for the TJMB highlands, although the lack of local paleobotanical data there complicates the estimating of local precipitation. The humid phases registered for other areas of the Levant during the Terminal Pleistocene and the Holocene also favored the highlands of Jordan.

Given the lack of local palynological data, scholars have proposed models of vegetation reconstruction based on indirect data. In particular, potential natural vegetation for the Levant, including Transjordan, has been proposed based on pollen records from the Hula basin and Lake Kinneret and on the basis of better-known regional patterns of climatic change (Van Zeist and Bottema 1991). This reconstruction is presented in maps of vegetation for three points in time (18,000–16,000, 12,000, and 5,000 years ago) included in *Tübinger Atlas des Vorderen Oriens* (*TAVO*;

Van Zeist and Bottema 1991). Although the attempt is valid in the scale and scope of landscape reconstruction in the Near East, it lacks detail in certain topographic features (valleys, depressions, slope orientation, etc.), soils, and humanized landscapes that should be evident, especially in the map pertaining to 5,000 years BP.

Gradual Woodland Reduction

The process of woodland reduction in the TJMB acts in conjunction with climatic deterioration and growing human pressure. As a result, the surrounding drylands increasingly advance at the expense of the woodlands and shrubland areas of the TJMB. The relatively continuous belt of wooded areas has been choked into small islands of wooded areas, as shown in figure 3.1. This process, often referred to as desertification, affects those areas on the borders of the Mediterranean type of climate in North Africa and Southwest Asia (Tomaselli 1981).

In the particular case of Jordan, desertification involves the transformation of wooded communities into shrublands and steppes and finally into desert (Feinbrun and Zohary 1955). Plant associations on the edges of the TJMB reflect the transitional character of plant communities as a function of precipitation and human impact across the Mediterranean and Irano-Turanian provinces (fig. 3.10). The thorny burnet garrigue (*Sarcopoterium spinosum*) dominates heavily grazed landscapes of Mediterranean climate in the highlands of Jordan. This type of garrigue rarely exists in areas with less than 300 millimeters of annual precipitation and is virtually absent from areas receiving less than 200 millimeters. In areas with annual precipitation between 200 and 300 millimeters, associations of *Salvia dominica* and *Ballota undulata* and of *Astragalus bethlemiticus* and *Marrubium libanoticum* replace the thorny burnet (fig. 3.10). Subsequently, in areas with precipitation below 200 millimeters a year, the *Ballota-Salvia* association is replaced by the more dryness-tolerant wormwood (*Artemisia herba-alba*) and anabasis (*Anabasis syriaca*) scrub, which are typical elements of the Irano-Turanian steppe. This example of plant community successions suggests that woodland reduction in the TJMB is more likely to have occurred as a result of human activities rather than following atmospheric desiccation.

In areas with less than 300 millimeters a year of precipitation, the process of woodland/shrubland reduction takes a different turn because moisture stress produces associations of scrub vegetation closely related

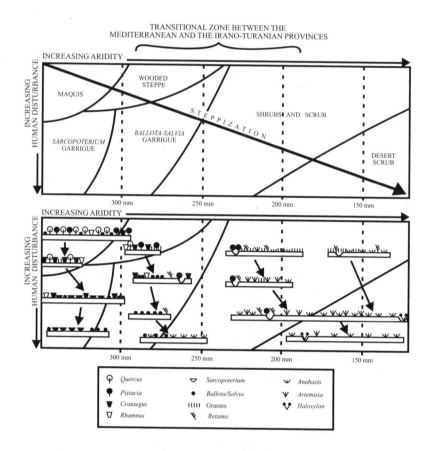

Figure 3.10 Steppization in relation to aridity and human disturbance as seen in transitional plant communities between the Mediterranean region and the Irano-Turanian steppe.

to the typical desert flora. The example shows that woodland degradation does not necessarily end in the invasion of desert plants, as the term *desertification* implies. Instead, a process of steppization, or the process by which a wooded area becomes a steppe, occurs on the edges of the Mediterranean climate.

Learning from a Relict Woodland

Testimonies of recent woodland reduction and steppization in Jordan are portrayed by relict tree stands located outside the main wooded areas,

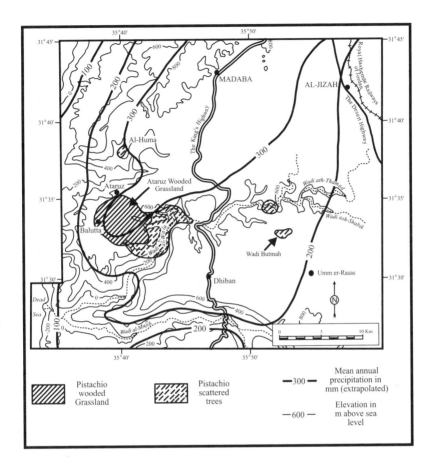

Figure 3.11 Present distribution of relict pistachio woodland communities in the Northern Moab Plateau (Madaba and Dhiban plateaus) in relation to elevation and precipitation.

such as the Ataruz wooded grassland (fig. 3.7). Located on the edge of the Northern Moab Plateau (fig. 3.11), the Ataruz wooded grassland is an isolated community of pistachio trees in the nonforest Mediterranean region. The scattered woods near Ataruz have been emphasized as a patch of relict woodland vegetation (Feinbrun and Zohary 1955; Bardtke 1956).

The earliest written and detailed account (Tristram 1873) of the woodland's existence derives from Henry Baker Tristram's visit to the nearby ruins of Ataruz and Machaerus (Mukawir). According to W. Amherst-Haynes, the botanist in Tristram's expedition, this was the only extensive wooded area in the entire territory of the Moab Plateau. In addition, two smaller spots with pistachio trees were spotted in the Wadi Butm (literally meaning "the valley of the pistachio tree") to the northwest of the ruins of Umm er-Rasas (Tristram 1873, 165; Wilson et al. 1881, 67). In the area described by these travelers, modern maps show only "Wadi Butmah," which bears a term that implies the presence of *butm* ("pistachio tree"). Two other locations with pistachio trees east of Dhiban have also been mentioned (Glueck 1939): edh-Dheheibeh (site 155) and Umm Sheijerat esh-Sherqiyeh (site 163). Although these trees could have been planted, it is possible that they are relicts of ancient vegetation. For instance, native trees in Jordan have commonly been preserved inside cemeteries out of respect for the dead (Al-Eisawi 1996).

In 2000, I carried out an extensive survey to identify and describe the pistachio relicts on the Northern Moab Plateau to verify whether they are the wooded sanctuaries described in the nineteenth century that have survived to this day. I determined that the Ataruz woodland and the communities in Wadi Butmah and other localities still exist (fig. 3.11).

Although the vegetation of Ataruz is similar to the one described in Tristram 1873, the extent of the wooded area may have been reduced. Amherst-Haynes (in Tristram 1873, 286) describes the area as a "park-like country," suggesting that the trees were scattered on a grassland area, much as they are today. All the wooded species described by the expedition exist today. *Pistacia atlantica* is the dominant tree species in the Ataruz area. However, today only mature pistachio trees persist; young trees are present only on cliffs, where they are out of the reach of goats. Among the shrubs described by Amherst-Haynes (Tristram 1873, 401–3), hawthorn (*Crataegus azarolus*) and buckthorn (*Rhamnus* spp.) exist today. Almond trees (*Amygdalus* sp.) were described as a wild component of the woods, but today they exist only in household orchards.

Oaks were not mentioned in Amherst-Haynes's botanical survey, and they are absent from the region today. It is possible, however, that oaks existed in this region in the recent past, as suggested by "Balutta," the name of a locality in the Ataruz area (fig. 3.11). *Balūt* is the Arabic word for both oak and acorn. There are no oaks in this locality, and its inhabitants do not remember seeing any, although they claim that the name refers to the existence of an oak tree several generations ago.

The absence of oaks in this area is a matter of further discussion, especially because in most Mediterranean wooded areas of Jordan and

Palestine, *Pistacia atlantica* is often associated with *Quercus calliprinos*. However, the Ataruz area receives only 200–300 millimeters of annual precipitation, which is marginal for oak development. Several researchers have pointed out that pistachio is more resistant to dryness than oak is (Feinbrun and Zohary 1955; Kürschner 1986; Baierle 1993; Al-Eisawi 1996). While *Quercus calliprinos* is a typical Mediterranean species, *Pistacia atlantica* is shared by the Mediterranean and Irano-Turanian provinces, which is evidence that pistachio trees are better adapted than oaks to dry steppe conditions. Furthermore, pistachio trees may also be more resistant than oaks to human pressure. Therefore, its resistance to dryness and human activities make the pistachio the last tree to be seen when traveling from the TJMB to the steppes (fig. 3.10).

The wooded steppe of Ataruz may have been a denser forest in the past, which would have required conditions of higher moisture and decreased pastoralism. Michael B. Rowton (1967) made a reference to forests north of the Arnon River (present-day Wadi al-Mujib), based on a passage in Flavius Josephus's *Jewish Wars*, in describing the siege of the Macherus fortress by Lucius Bassus: "Having settled affairs here, Bassus pushed on with his troops to the forest called Jardes, it being reported that many who had previously fled from Jerusalem and Macherus during the respective sieges had congregated in this quarter . . . he then ordered the infantry to fell the trees among which the fugitives had taken cover" (Josephus 1968, Book VII, chap. VI (5) 565). Whether this forest was situated near the site of Macherus (Mukawir) or somewhere else farther north, perhaps in the northern highlands, is unclear. A translator's note at the foot of the text refers to Jardes as an unidentified site, possibly a corruption of "Yarden" (Hebrew for the Jordan River).

The passage implies the presence of a wooded area dense enough to provide cover to the fugitives, suggesting a relatively dense community. The event described had occurred at the end of the first century AD, which is part of a relatively humid phase in the region, according to paleolake-level data in the Dead Sea (Frumkin et al. 1994; Frumkin and Elitzur 2002). Nonetheless, evidence for a forest in Ataruz on the basis of the written passage is inconclusive.

Examples of wooded steppes with pistachio trees similar to Ataruz have been reported for Jordan in the area east of Jerash and Irbid and in areas in the Golan and the Hauran (Schumacher, Oliphant, and Le Strange 1886), as well as on the western edges of the southern highlands (Bardtke 1956; Baierle 1993). Perhaps these relict wooded communities are evidence with which to reconstruct the once extensive woodland areas of preagricultural times.

Mediterranean Agriculture on the Edge

The modern distribution of wild cereals along the Mediterranean woodlands of the Near East led many scholars to view this region as the cradle of cereal domestication (Zohary 1969; Köhler-Rollefson 1988; Henry 1989). Thus, cereal domestication took place on an area of open woodlands and wooded grasslands supported by red Mediterranean soils.

The region was also the domestication center for other typical Mediterranean crops such as olives, legumes, and grapes, which figure among the remains retrieved from Neolithic and Chalcolithic sites (Neef 1997). Although a number of crops have been introduced and nontraditional agricultural forms have been implemented in this region, traditional agriculture there is still based on the cultivation of wheat, barley, olives, legumes, and grapes. But how similar to the agriculture of the areas around the Mediterranean Sea is agriculture in the inland and somewhat dry environment of Transjordan?

The four basic elements of Mediterranean agriculture proposed by David B. Grigg (1974) are present in Jordan: (1) the predominance of cereal cultivation, in particular wheat; (2) livestock grazing, especially goats and sheep; (3) horticulture, especially olives and grapes; and (4) cultivation of fruits and vegetables. Owing to the marginal character of rain-fed agriculture in Jordan, elements 3 and 4 have lesser importance than in other areas of Mediterranean climate in the Levant. Toward the Irano-Turanian steppes, elements 3 and 4 disappear, while element 1 decreases and element 2 increases.

The location of the TJMB at the transition to drier climates brings to the fore the issue of the Mediterranean agriculture frontier, which is visible when looking at the eastern limits of wheat, barley, and olive cultivation today (fig. 3.12). In areas to the west, toward the drylands of the Jordan Valley where irrigation projects have been implemented, the agricultural boundary is diffuse. But to the east, the distribution of these crops is a direct response to the decline of precipitation (fig. 3.12).

The eastern boundary of the rain-fed olive tree cultivation can be used to define the limits of Jordan's Mediterranean climate. However, this assumption should be made carefully, since irrigation in some areas has opened the possibilities of expansion of most Mediterranean crops into

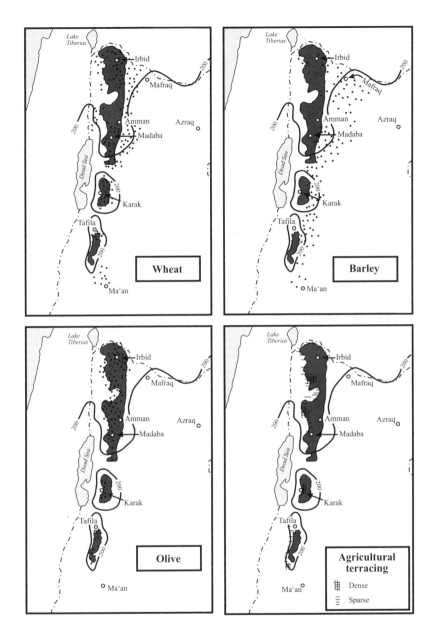

Figure 3.12 Present distribution of wheat, barley, and olive cultivation and agricultural terracing. The extent of rain-fed irrigation roughly coincides with the isohyet marking 200 millimeters of average annual precipitation, with rainfall greater toward the west. The dark areas mark the distribution of red Mediterranean soils. (From Royal Jordanian Geographic Centre 1984:117 and 123, with modifications by the author.)

drier areas. Olive tree cultivation concentrates on the steeper slopes on the plateau margins and the mountains where terracing has been implemented.

The eastern boundaries of barley and wheat vary in relation to their slightly different requirements in terms of water and soil. Wheat needs higher precipitation (above 200 millimeters a year) and loamy-clay, deep, well-drained soils (Beaumont and Atkinson 1969). These characteristics are found in the area of red Mediterranean soils but not on the yellow soils of the steppe or on the stony soils on steep slopes of the mountains. Barley is less demanding, for it can grow in stony, thinner soils, under lower amounts of precipitation. Furthermore, barley is more tolerant of alkaline and saline soils. This makes barley one of the crops commonly farmed by modern seminomadic Bedouins in wadi bottoms and alluvial fans in those areas of the desert receiving less than 100 millimeters of rain a year. This type of risky and occasional agriculture reflects the transitional character of Mediterranean agriculture into the desert and the strategies that part-time nomadic groups develop to increase food production.

Agricultural terracing, a technique developed to control erosion and to increase the areal extent of agricultural surfaces on steep slopes, is an aspect of Mediterranean agriculture also present in the TJMB. Although it is not as widespread as in other mountainous areas of the Levant, terraced landscapes are common in some areas on western slopes of the TJMB facing the Jordan Valley (fig. 3.12). The density of terraced lands is more common in the north, especially in the stretch of steep land from Salt to Wadi al-ʿArab, including the Ajlun Mountains. South of Salt, terracing is more scattered, except for the area around the town of Karak. Although olive cultivation dominates the terraced slopes, very often fruit and nut trees are planted, in some cases alternating with olive trees.

Although cereals and olives seem to dominate the landscape of traditional agriculture, legumes, grapes, and orchard vegetables play an important role in family-run cultivation. Legumes and grapes are cultivated on a smaller scale, especially in family gardens. Grape cultivation has low representation in today's Mediterranean agriculture in Jordan, partly because wine production is not important in a Muslim country such as Jordan. Grapevines are, however, typical in home gardens, since their leaves and fruits are eaten. In centuries previous to the Islamic era, grapevine production, associated with wine making, reached important levels among the agricultural products of the region. Plant macroremains recovered from sites in the highlands of Jordan show that grape production peaked during the Roman and Byzantine periods (Gilliland 1986). The list of vegetables is long and varies with the season. Carol Palmer (1998) gives a detailed description of crops and their yearly schedule as well as the different methods of plowing and fallow systems. Her study is based on traditional farming on the Irbid Plateau and the northern slopes of the Ajlun Mountains.

One important aspect of the cultural landscape in the TJMB is the abundance of field weeds, known collectively as segetal vegetation. The TJMB is highly rich in weeds, most of which are common in other Mediterranean regions of the Levant. The number of species of weeds in the TJMB is immense, although most of them belong to a few families: Asteraceae, Boraginaceae, Caryophillaceae, Convolvulaceae, Cruciferae, Labiatae, Fabaceae, Geraniaceae, Liliaceae, Papaveraceae, Polygonaceae, and Scrophulariaceae. Of these, the largest number of species belongs to the Asteraceae, which dominate most of the fallow fields, roadsides, and disturbed areas. The most abundant genera of Asteraceae are *Anthemis, Carlina, Carthamus, Centaurea, Cynara, Evax, Notobasis, Onopordon, Picnomon, Scolymus, Silybum*, and *Varthemia*. Other families of weeds include mainly one species, as is the case of the Euphorbiaceae, which includes *Euphorbia hyerosolimitana*, one of the most common weeds in the fields of the Jordanian highlands. The importance of weeds is often not acknowledged when interpreting pollen and other paleobotanical records. However, they seem to be an important part of agricultural life, which has dominated the Middle East for the past nine thousand years.

Recent Developments

The reduction of the Mediterranean arboreal communities in Jordan has been profound, especially in areas with potential for agriculture. Because little is known about the degradation of forests in the past, the blame for most of the destruction of wooded areas in Jordan and Palestine has fallen on the Ottoman administration (see Lowdermilk 1944; Chapman 1947; Reifenberg 1955). Such a belief is founded on the descriptions made by Western travelers of the numerous wooded areas in the country in the nineteenth century (Bardtke 1956; Atkinson and Beaumont 1971; Abujaber 1989).

Demand for timber for the construction of the Hejaz Railway in the early 1900s was the last blow that reduced the southern woodlands to a steppe of scattered trees. Because most of the railroad line went through steppe, mainly along the former Hajj Road (fig. 3.1), timber had to be obtained in the nearest woodlands. In the case of the southern highlands, a subsidiary line was constructed from Qaʿala Anayza, on the main railroad line, north of Maʿan, to Hisha, near the wooded area (Pick 1990).

Although the Ottoman administration is to blame for part of the destruction, the reduction of the arboreal communities has been a gradual process occurring over millennia. Unfortunately, there are no written records for years previous to the last two centuries of Ottoman administration.

Ironically, despite the rapid growth of population in the past fifty years, the area covered with trees has grown in Jordan as a result of numerous reforestation projects. Thus, the nonforest Mediterranean vegetation region looks more forested today. With the exception of Aleppo pine, most of the species of these reforested areas are not native to Jordan. Eucalyptus (*Eucalyptus* spp.), casuarinas (*Casuarina* spp.), acacia (*Acacia cyanophila*), and other drought-resistant species figure among the most common introduced trees in Jordan (Hatough-Bouran et al. 1998). Although the increase in forested areas is welcomed by the people of Jordan, the widespread occurrence of reforested areas poses a problem for determining the presence of relict native trees. Therefore, the more foreign trees that are planted, the more difficult spotting relicts of woodlands in the landscape becomes.

4
Encroaching Drylands
Steppe and Desert

The Arabic language is rich in terms referring to desert landscape features—so rich, indeed, that many names are now used as standard terms in arid land geomorphology (e.g., *barchan, seif, sebkha,* and *wadi*). In Jordan alone, the variety of names referring to specific qualities of steppe and desert landscapes is vast.

Although *ṣaḥrā*, the generic and formal term for deserts, is often used in road signs and maps, the most common terms used in Jordan speech are *bādia* (or *bādiyah*) and *barrīyah*. The term *bādiyah*, from which the word *bādawī* ("Bedouin") derives, refers to the nomadic way of life practiced in drylands. Therefore, it is used to denote not only deserts but also steppes, that is, the Irano-Turanian steppe. The term *barrīyah* derives from *barrā* ("outside"), a term that defines the desert as an area outside of cities and villages, namely, where people carry on a nomadic way of life. Thus, *barrīyah* denotes the cultural aspect of deserts rather than a physical landscape. The Arabic word for steppe is *suhb*, which is a word in the same formal category as *ṣaḥrā*. The term *marj*, common in the colloquial speech of Syria and Jordan, refers to a rather grassy steppe, often located in areas with more water.

In Jordan, the types of deserts are defined in a variety of terms in relation to their particular physical characteristics. Among the most used in Jordan are *ḥārrah* (stony desert on basalt), *hamaḍ* (stony desert on limestone), and *ramlah* (sandy desert).

The drylands of Jordan occupy approximately 90 percent of the territory and include the Irano-Turanian steppe, the Saharo-Arabian desert, and the Sudanian dry savanna (see fig. P.1, preface). The diversity of landforms and lithology across the Jordanian landscape provides each of these regions with regional variants. In the Saharo-Arabian desert, for example, one can find three main desert types: the black basaltic deserts of northeastern Jordan (the Eastern Badia), the limestone deserts of Wadi Sirhan and the area of Al-Jafr, and the sandstone and sand-dune desert landscapes of southern Jordan. In addition, each of these regions presents a variety of meso- and microenvironments, depending on the presence of water or smaller landform units. In the Eastern Badia, for example, a great distinction is often made between the basalt (hārrah) and limestone desert (hamaḍ). Indeed, the two designations are used as proper names for these two regions of northeastern Jordan (fig. 4.1-A).

Landforms and lithology are not the only aspects that define desert landscapes. Climatic designations such as semiarid, arid, and hyperarid are commonly used in the geographic literature, particularly when referring to precipitation. Additionally, vegetation types are often used in most designations of arid and semiarid landscapes. Generically, these designations include steppe, wooded steppe, desert grassland, desert shrubland, desert scrubland, alkali scrub, and so forth. Besides providing information on plant life forms, designations of vegetation types are useful in cultural ecology and archaeology because they relate to the interaction between humans and the environment.

Diversity in an Apparently Monotonous Landscape

The variety of landforms in the deserts of Jordan varies, depending on lithology, structure, and tectonics. Inselbergs, alluvial fans, sand dunes, and playas are common in the predominantly sandstone territory of the Hisma basin and Wadi Rum, while extensive alluvial fans, sand dunes, playas, and craggy hills are typical of the granite territory of the southern Wadi Araba. Basalt mesas and extensive stone surfaces characterize the Black Desert in the northeast of the country. The loess plains on the fringe of the desert are concentrated in several regions, most notably the Mafraq and Khan az-Zabib plateaus and the Wadi al-Hisma. This geological and geomorphological diversity influences the spatial arrangement of plant associations, since some species are directly associated with substrate properties (e.g., sandy sediments, saline soils, carbonated rocks, alluvial fan gravels).

The diversity of plant and animal species in the steppes and deserts of Jordan is the result of an intricate combination of factors such as local climate, ecological niches, and anthropogenic effects. Therefore, changes in vegetation are expected to be seen even within a small area.

During the agricultural and pastoral period (the past 9,000 years), the diversity of fauna in the Levant and in the Near East has been dramatically reduced. Consequently, animal diversity in Jordan is low compared

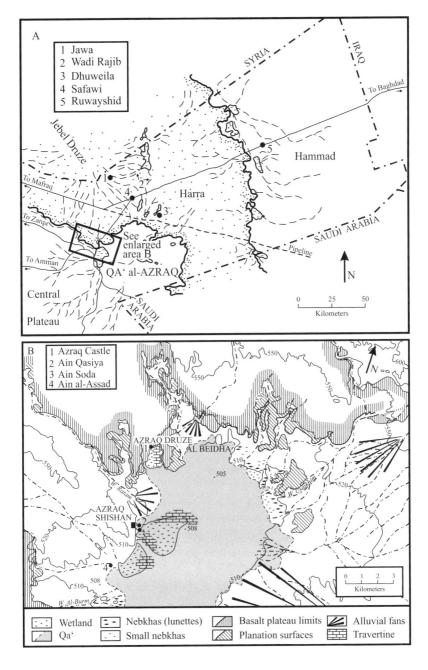

Figure 4.1 (A) Distribution of major landform units within the Badia region, and box indicating the Azraq area (based on Betts 1998:figs. 2 and 3, with modifications by the author). (B) Distribution of landforms in the Azraq area (based on Besançon, Geyer, and Sanlaville 1989:fig. 7, with modifications by the author).

Figure 4.2 *Artemisia herba-alba* within the Irano-Turanian steppe on limestone substrate near Wadi ar-Rumeil.

to other drylands of the world, where human impact is a more recent phenomenon.

Plant Groups and Their Ecological Implications

It is difficult to determine how many of the country's 2,500 species of vascular plants are exclusively steppe and desert flora, because some plants are shared with the Mediterranean region. However, certain groups of plants are found almost exclusively in the steppes and deserts. They include mainly sages and chenopods. Grasses are in theory the original vegetation of the steppes, despite their low representation in such areas today. Other groups of dryland plants include scrub and shrubs belonging to different families. Therefore, given the tendency of some plant groups to dominate in the drylands, analyzing their main components and ecological significance is important.

Sages

Typical for their distinctive scent, sages (*Artemisia* spp.) are the most abundant and typical plants of the steppes and deserts of Jordan (fig. 4.2). Three species of *Artemisia* characterize the dry landscapes of Jordan. The most common is *Artemisia herba-alba*, a perennial in which the living parts of the aerial shoot become reduced in summer, allowing propagation by both seeds and division of the subterranean stem into several shoots (Zohary 1973). Known to the Bedouins as *shīḥ*, *A. herba-alba* has several uses, from medicinal ones to fuel. But because it is avoided by livestock, it is one of the most common antipastoral plants in the steppes and deserts. Other species of sages in Jordan are *A. judaica* and *A. monosperma*.

The incidence of *Artemisia* in pollen diagrams of the eastern Mediterranean region is often interpreted as a signature for periods of aridity. However, compared with the chenopods, *Artemisia* reflects less extreme aridity (El-Moslimanny 1990; Rossignol-Strick 1995). Despite its association with the effects of pastoralism, the presence of *Artemisia* in pollen diagrams is seldom equated with pastoral activities.

Chenopods

The family Chenopodiaceae (the goosefoot family) has a very wide distribution in the world's drylands because of chenopods' great tolerance for dryness, high soil pH, and high salinity. Although the Chenopodiaceae thrive in arid lands, they are often found as ruderal plants in disturbed areas elsewhere.

The entire spectrum of chenopods in the Jordanian drylands is reflected in the ecological affinities of each species. *Haloxylon persicum* and *Hammada scoparia* are typical of sandy deserts and wadi bottoms of the Saharo-Arabian desert. *Anabasis articulata*, *A. syriaca*, and *Noaea mucronata* dominate the landscape of the Irano-Turanian steppe. Other species of the genera *Atriplex* and *Suaeda* are adapted to saline soils.

Unfortunately, pollen diagrams offer little information about the distribution of chenopods during prehistoric and historic periods. Part of the problem is that chenopods in pollen diagrams are abundant not because of their existence nearby but because they are brought into the region by dust storms from desert areas (Weinstein 1979). In some cases, such storms can bring pollen grains from regions as far away as North Africa. Additionally, pollen diagrams do not show lower taxonomic groups (i.e., genus and species) within the chenopods. These difficulties imply that the ecological history of the local chenopods cannot be reconstructed.

Grasses

The family Gramineae (or Poaceae) dominates in steppes and savanna vegetation. In the Middle East, grasslands are limited to areas receiving more than 200 millimeters of annual precipitation in the highlands of Syria and the Anatolian Plateau (Zohary 1973). In drier regions, grasslands are limited to those areas where special conditions for dominance of grasses exist, including, for example, a high water table or constant flooding, which provides higher levels of moisture and nutrients. Thus, in the most humid areas of the Irano-Turanian steppe in Jordan, the amount of precipitation (200–300 millimeters) would be enough to maintain a significant grass cover. However, livestock grazing has occurred for so long that discerning whether the natural vegetation was grass or sagebrush becomes difficult (Rossignol-Strick 1995).

Numerous archaeobotanical studies suggest that the dominance of grasses in the semiarid and arid Near East has diminished since the beginnings of agriculture and pastoralism (Hillman 1996; Miller 1997). Therefore, the share of grasses in the modern Irano-Turanian steppe today is low compared to the dominant sage and scrub vegetation (e.g., *Artemisia*, plants of the family Labiatae, and various chenopods). In a study comparing plant diversity in grazed and nongrazed plots in the Irano-Turanian steppe, a highly diverse set of grass genera was present in the nongrazed areas (Hatough-Bouran, Al-Eisawi, and Disi 1986). Therefore, it is logical to believe that grasses had a strong presence in the steppe environments in times preceding the era of grazing by domesticated ungulates.

The distribution of grasslands in prehistoric times has always been a topic of interest among scholars studying the origins of agriculture, particularly because wild cereals were an important component in the grassland ecosystems of the Middle East at the end of the Pleistocene (Zohary 1969; Hillman 1996). One possibility is that most wild grasses occurred along the transition between the Mediterranean and the Irano-Turanian region (Zohary 1969). The intense transformation of grasslands in subsequent millennia makes it difficult to delineate the specific ecological niches that wild cereals occupied, but the abundant recurrence of wild and domesticated grasses in other areas suggests that wild cereals may have occurred in other areas of the steppe as well (Hillman 1996).

A pollen diagram from Birkat Ram (Golan Heights) illustrates that the frequencies of grasses are relatively low for times predating the

Persian-Babylonian period (Schwab et al. 2004). Frequencies reach a peak during the Hellenistic and Roman periods and begin to decline during the Byzantine period, to reach a minimum during the Early Islamic, Crusader, and Mameluke periods. In more recent centuries, grass pollen frequencies increase again but do not reach the level previously attained during the Roman-Hellenistic periods. Why grasses became more abundant during this period is unclear. The introduction of Hellenistic and Roman economies may have resulted in less grazing (Schwab et al. 2004). As archaeobotanical data indicate, villages and towns grew during these periods, suggesting that rural economies were primarily focused on farming, especially through the production of olive, grapevine, and cereals (LaBianca 1990). The Birkat Ram diagram is located on the ecotone between the nonforest Mediterranean region and the Irano-Turanian steppe. Therefore, it is a good example of human-induced changes in areas that formerly supported more grasses than today.

Other Herbs, Scrub, and Short Shrubs

In addition to the typically abundant chenopods and sages, many species of herbs occur in the associations of the steppes and deserts. In most cases, these herbs are associated with specific lithology or with a certain type of land use. Plants of the family Asteraceae are very abundant throughout Jordan. The Asteraceae is a very diverse family with a high number of genera and species. The Liliaceae is another family with a relatively strong presence in the drylands of Jordan. Some of the most representative members of the Liliaceae are *Urginea maritima* and other monocots, such as the genera *Asphodelus, Bellevalia, Colchicum*, and *Ornithogalum*.

    Zygophyllum dumosum is a Sudanian shrub element that is widespread in the steppes and deserts of southern Jordan and the Rift Valley. *Calligonum comosum* is a typical species of sand-dune environments, where it is commonly found in association with *Ephedra transitoria* and *Haloxylon persicum*. Other plants seem to be adapted to sandy and gravelly grounds and to conditions of high salinization. *Onobrychis hemicycla* and *Alhagi marorum*, members of the family Fabaceae, are very abundant in the drylands of Jordan. Besides their high tolerance for salinity, they abound in areas of intense grazing. Likewise, *Capparis spinosa* and *Peganum harmala* are plants found everywhere in the drylands of Jordan because of their antipastoral characteristics.

Trees and Tall Shrubs

Although the idea of steppe and desert discards by default the presence of trees, the drylands of Jordan have niches where relict tree stands are found. Almond (*Amygdalus korschinskyi*) and pistachio (*Pistacia atlantica*) are two species present in isolated locations in the deserts, where they exist as relict stands (Baierle 1993). Relict communities of pistachio trees are also found in numerous locations in the deserts of Syria (Willcox 1999), Jordan (Al-Eisawi 1996), the Negev, and the Sinai (Danin 1983). These relict stands are often found along wadis, as is the case of Wadi al-Butm in the vicinity of Qasr al-Amra on the Central Plateau. Other shrubs of Mediterranean origin frequently found in the steppes and deserts are *Crataegus aronia* and *Rhamnus palaestinus*, which are tolerant of dryness and moderately resistant to overgrazing.

The varieties of *Acacia tortilis* are the most widespread trees in the Sudanian region of Jordan, where the species forms associations with other typical trees and shrubs of the African savanna (e.g., *Moringa peregrina*, *Ziziphus spina-christi*, and *Loranthus acaciae*).

Nerium oleander, a typical eastern Mediterranean shrub, is commonly found in wadi bottoms in the Irano-Turanian steppe. It is rarely found in the Sudanian and Saharo-Arabian deserts. *Retama raetam*, another widespread shrub, is shared by the Mediterranean, Irano-Turanian, and Saharo-Arabian provinces. In extremely dry areas, this shrub is found in wadi bottoms, where it could be mistaken for *Tamarix*, another species common in desert wadi bottoms.

In the past, trees were common around springs, but the frequent visits of goats and sheep to watering holes have diminished their coverage. Trees are more likely to be found in places where topography keeps them from the reach of goats and humans. Such is the case for *Ficus pseudosycamorus* (wild fig tree) at the springs along cliffs in the Wadi Rum area (fig. 4.3-A). *Phoenix dactylifera* (date palm) is common at springs, but its presence at these locations is likely the result of human intervention.

The Irano-Turanian Steppe

The Irano-Turanian vegetation territory surrounds the Transjordanian Mediterranean Belt in all directions within an area that receives between 250 and 100 millimeters of annual rainfall (Al-Eisawi 1996). In general terms, the Irano-Turanian steppe region is the transitional zone between

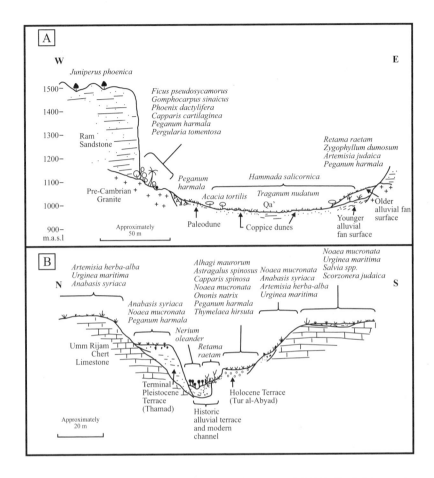

Figure 4.3 Vegetation transects in (A) Wadi Rum and (B) Wadi ath-Thamad.

the areas with Mediterranean climate and the deserts. Also described as dwarf-shrublands (Frey and Kürschner 1989), this region covers a wide area in the Middle East, forming an arch parallel to the Fertile Crescent from Jordan to Iran, extending across Syria, eastern Turkey, and northern Iraq. This broad geographic range gives the Irano-Turanian steppe a high variety of vegetation formations, of which the three most important are *Artemisia herba-alba*, *Anabasis haussknechtii–Poa sinaica*, and *Haloxylon articulatum–Salsola villosa* (Feinbrun and Zohary 1955).

Associations of *Artemisia herba-alba* are widespread on rocky substrates, mainly in the driest parts of the territory. Associations of *Anaba-*

sis haussknechtii and *Poa sinaica* are common on areas of loess deposits, alluvial terraces, and colluvial deposits, especially areas that have been heavily grazed. The associations of *Haloxylon articulatum* and *Salsola villosa* are commonly located in wadi bottoms and on saline soils.

Other important species in the Irano-Turanian territory of Jordan include *Noaea mucronata, Astragalus spinosus, Urginea maritima,* and *Asphodelus aestivus* (fig. 4.3-B). Among shrubs and trees, *Retama raetam* and *Tamarix* spp. are perhaps the most common. On the slopes of the highlands facing the Jordan Valley, the Irano-Turanian steppe elements intermingle with elements of Sudanian vegetation such as *Ziziphus lotus* and *Z. spina-christi.*

The Irano-Turanian steppe vegetation has been heavily grazed and in some places completely removed through farming (Feinbrun and Zohary 1955). This process was so intense that soils were eroded, hence creating an irreversible process for the recolonization of former vegetation. Consequently, the Irano-Turanian steppe bears a strong resemblance to the surrounding deserts. Look for example to the landscape along the road between Zarqa' and Azraq. In other areas along the Desert Highway, which in fact runs along the Irano-Turanian steppe, one can see only barren limestone and patches of loess bearing only patches of *Anabasis* scrub.

The Saharo-Arabian Desert

This region is widely distributed across the southern part of Southwest Asia. It includes the driest regions of the Levant, southern Iraq, and the Arabian Peninsula. The Saharo-Arabian region occupies approximately two-thirds of the territory of Jordan (fig. P.1, preface). It is, however, underpopulated and with less rural and urban development than the other three landscape regions of Jordan.

The limits of the Saharo-Arabian region with the adjacent Irano-Turanian and Sudanian regions are difficult to trace because vegetation transitions are gradual, with plenty of ecological indentations (Zohary 1973). Therefore, the boundaries of the Saharo-Arabian deserts are primarily determined by climatic rather than ecological criteria. The boundary with the Irano-Turanian region runs along a line of slightly under 100 millimeters of annual precipitation. The boundary with the Sudanian territory is defined by the 10°C isotherm for January average temperature. The Sudanian region is, for the most part, frost free, which permits the growth of tropical floristic elements. Despite this temperature limit, a variety of plant species of Sudanian affiliation thrive in niches deep inside the Saharo-Arabian region (Boulos 1977).

The Saharo-Arabian region in the Levant is relatively poor in species, and the bulk of the territory is occupied by associations dominated by *Anabasis* spp. and *Retama raetam* (Zohary 1973). *Anabasis* prefers rocky terrains, while the latter is often found on sandy ground. A larger variety of plant species exists in the wadi bottoms, where deeper soils and adequate moisture are available (Al-Eisawi 1985). These niches also hold relict communities of trees and shrubs of the Mediterranean, Irano-Turanian, and Sudanian regions. Two examples of Mediterranean/Irano-Turanian vegetation are pistachio and almond, which are commonly found in wadis (Al-Eisawi 1996), a pattern also observed in the Negev and the Sinai (Danin 1983).

In the Saharo-Arabian territory of Jordan, plant species and their associations vary depending on the types of substrate, whether gravel, runoff hammada, sandy hammada, or basaltic pebbles (harra) (Al-Eisawi 1996).

The runoff hammadas include the bottoms of most wadis and areas subject to flash flooding. The terrain is usually composed of gravel and sand, but areas with silt and clay occur in some places. The main species found on these grounds are *Tamarix* spp., *Retama raetam, Acacia tortilis, Artemisia judaica*, and *A. monosperma*. Trees such as *Pistacia atlantica* and *Amygdalus arabica* are found in a few valleys where microclimatic conditions allow moisture storage.

The sandy hammada corresponds mainly to sand dune fields and alluvium. Common species on the sandy hammada include *Seidlitzia rosmarinus, Capparis leucophylla, Ephedra transitoria*, and *Calligonum tetrapterum*, among others. Gravelly surfaces occupy extensive areas of the Saharo-Arabian deserts of Jordan. The prevailing species of this type of substrate is *Seidlitzia rosmarinus*, which is a low shrub that forms associations with other herbs. *Spergularia diandra, Anthemis desertis, Gymnarrhena micrantha*, and *Stipa capensis* are among the common plants on gravelly surfaces. The pebble hammada, or harra, is a stony and gravelly terrain of basalt, where *Salsola vermiculata* and *Halogeton alopecuroides* predominate.

It is also important to include saline vegetation that covers playas (*qaᶜ*). The most important halophytes growing on these grounds are *Nitraria retusa, Alhagi maurorum*, and various species of *Tamarix, Suaeda,* and *Juncus.*

The Sudanian Vegetation

The Sudanian floristic territory in the Levant is interpreted as a deep wedge reaching northward along the Rift Valley and extending to the 32nd parallel (Zohary 1973). This wedge is discontinuously connected to the core of the Sudanian vegetation of the eastern Sahel in North Africa, the southern third of the Arabian Peninsula, southeastern Iran, the lower valley of the Hindus River in Pakistan, and the western end of the Deccan Plateau in India (Shmida and Aronson 1986). Subsequently, names such as Sudano-Sindian, Nubo-Sindian, and Sudano-Deccanian have been applied to this region.

In Jordan, the Sudanian floristic province extends mostly along the Wadi Araba and the shores of the Dead Sea basin and into the Jordan Valley. Another extension of the Sudanian realm penetrates the territory of Jordan into the sandstone region of Wadi Rum and Qaᶜ Disa. Although the geographical distribution of the Sudanian territory in the Levant seems like a continuous extension of African tropical vegetation, in reality the Sudanian province in the southern Levant is formed by scattered enclaves (Zohary 1973). In Jordan, these enclaves include mostly the bottom of some of the canyons (e.g., Wadi al-Mujib, Wadi al-Karak, and Wadi al-Hasa), some of the valleys northwest of Aqaba, and areas around springs. The rest of the territory is a mixture of Sudanian vegetation with elements of other provinces, mainly the Irano-Turanian and Saharo-Arabian regions.

Sudanian elements penetrate these regions, as is the case of *Ziziphus spina-christi*, which intermingles with Irano-Turanian elements in the Jordan Valley and along the escarpment bordering the plateaus. *Acacia tortilis* subsp. *raddiana* penetrates into the Saharo-Arabian region in the sandstone area (e.g., Wadi Rum and Qaᶜ Disa) and alluvium derived from the Crystalline Mountains of southern Jordan (fig. 4.4).

According to Michael Zohary's (1973) map titled "Geobotanical Outline of the Middle East," another wedge of Sudanian vegetation penetrates the Saharo-Arabian territory from the Arabian Peninsula along the Wadi Sirhan depression. This territory, however, has been assigned to the Saharo-Arabian region by Dawud al-Eisawi (1985, 1996).

Figure 4.4 Acacias on an alluvial fan in the southern Wadi Araba.

As mentioned above, the limits between the Saharo-Arabian and Sudanian regions are difficult to trace because of the gradual nature of their boundaries. The best way to view this limit is through the bioclimatic classification in Long 1957, modified by Al-Eisawi (1985) (see fig. 2.10 and table 2.3). Accordingly, the bulk of the Sudanian territory falls within the "very warm variety of the Saharo-Mediterranean Bioclimatic Zone" (Al-Eisawi 1985). In this zone, the mean minimum temperature ranges between 8°C and 12°C, while absolute minimum temperatures rarely fall below 3°C. Therefore, the main limitations for the northward migration of Sudanian flora are lower winter temperatures and lack of summer rains (Shmida and Aronson 1986). But how can the presence of this type of vegetation reaching latitudes over 30° N be explained?

The answer to this question may be exhibited by the modern landscape, especially in (1) the distribution of enclaves of Sudanian vegetation inside other vegetation regions and (2) climate changes that favored penetration of Sudanian vegetation into the Levant. The isolated enclaves of Sudanian vegetation are found in canyons and in areas of the Central Plateau. Numerous species of the Sudanian vegetation have been reported in Al-Jafr basin (Boulos 1977), a territory deep inside the Saharo-Arabian floristic region. The location of Sudanian enclaves in other vegetation regions corresponds to local microclimatic conditions that allow tropical plants to endure harsh winter temperatures. In the canyons, for example, Sudanian elements are protected from frosts. Many woody Sudanian elements are concentrated on or around alluvial fans, where groundwater is commonly found in layers. Some of the Sudanian shrubs and trees (e.g., *Acacia* and *Ziziphus*) develop deep root systems that can access such layers and tap water for use during the long dry periods. This source of water may make up for the lack of summer rain.

Past climate changes played a significant role in the formation of Sudanian vegetation enclaves in the interior of Jordan. The expansion of Sudanian floristic elements would have benefited from conditions of warmer temperatures and increased summer rain, which evidently occurred in the Levant at certain times during the Upper Pleistocene and Early Holocene (Bar-Matthews, Ayalon, and Kaufman 2000). However, opinions of scholars working on this issue differ with regard to the timing for the arrival of Sudanian flora in the Levant (Shmida and Aronson 1986). Faunal remains from early Pleistocene deposits in the Jordan Valley indicate that northward migration of fauna occurred throughout the Tertiary and during the warm stages of the Pleistocene (Tchernov 1988). However, there are no plant remains of Miocene, Pliocene, and early Pleistocene age that support a northward migration of plants. Furthermore, the low number of endemisms in the Sudanian flora of the Levant indicates that the alleged migration from the south occurred relatively recently, most likely after the last glacial period (Shmida and Aronson 1986). This argument is compelling, considering that a decrease in temperatures and rainfall during the glacial maxima may have created conditions completely unfavorable to plants of tropical origin. Although Avi Shmida and James Aronson (1986) point to a massive immigration of Sudanian flora into the rift between 8 ka and 4 ka BP, when conditions were warmer and more humid, the migration may instead have initiated during the Terminal Pleistocene, especially during the warm and wet periods between 15 ka and 11 ka BP (see chapter 6). In addition to warmer and wetter conditions, these periods were characterized by the northward expansion of the Indian Ocean Monsoon, which brought summer rains to the southern Levant (Bar-Matthews, Ayalon, and Kaufman 2000).

The possible paths of northward migration of Sudanian elements into the Levant may have been along the Rift Valley, the Wadi al-Hisma and Wadi Rum of southern Jordan, and the Wadi Sirhan depression

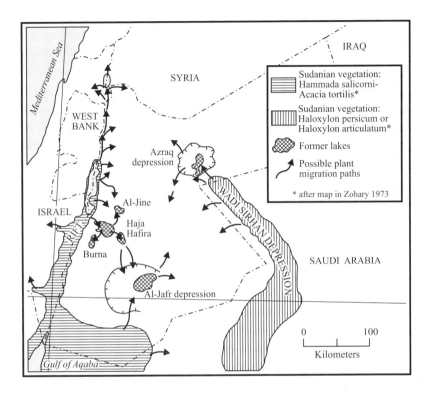

Figure 4.5 Hypothetical pathways of northward migration of Sudanian vegetation during the warm and humid stages of the Pleistocene and Early Holocene.

(fig. 4.5). The canyons dissecting the plateaus (e.g., Wadi al-Mujib and Wadi al-Hasa) may have served as secondary corridors for the intrusion of Sudanian flora as well. Today these canyons present large amounts of Sudanian floristic elements. Other secondary pathways include the lacustrine basins of the Central Plateau, where moisture and temperature conditions were perhaps favorable. In some cases, some of the relict species have disappeared from the Sudanian realm in the Levant, as is the case of *Acacia nilotica* (mainly from overexploitation) (Kislev 1990).

Faunal remains for the Middle and Upper Pleistocene in Azraq point to distinctive African steppe and savanna species (Rollefson et al. 1997; Rollefson 2000). Floral migrations may have accompanied these northward movements of fauna. However, until detailed paleobotanical studies prove this, it will not be possible to confirm the recent migration scenario for vegetation. The use of carbon isotopes and phytoliths from paleosols may provide some information that could help track down the past expansion and retraction of Sudanian vegetation in the modern territory of Jordan, as has been done in the Arabian Peninsula (Ishida et al. 2003; Parker et al. 2004).

The most common associations of Sudanian vegetation include those dominated by *Acacia* spp., which forms enclaves confined to the vicinity of permanent springs or larger outlets of water (Feinbrun and Zohary 1955). The bases of the alluvial fans along the Araba Valley and the Wadi Rum areas are typical localities where acacias dominate the arboreal vegetation. *Hammada salicornica* is a shrub that predominates on stony pavement, although it can be found elsewhere. Farther north, along the shores of the Dead Sea and along the Jordan Valley, the enclaves are dominated by *Ziziphus spina-christi* (Feinbrun and Zohary 1955), marking the transition between the Sudanian and Irano-Turanian regions in the Rift Valley.

The most common species growing on sand dunes are *Haloxylon persicum, Calligonum comosum, Plantago ovata, P. cylindrica, Silene villosa, Panicum turgidum*, and *Nerudada procumbens*, among others. Al-Eisawi (1996) has observed that dune fixatives in the Wadi Araba include *Retama raetam, Calligonum comosum*, and *Haloxylon persicum*. When the latter reaches heights between three and four meters and diameters larger than five meters, it is an indication that vegetation has reached a climax (Al-Eisawi 1996). This can be a good bio-indicator of sand-dune stabilization, which is useful when interpreting development of sand dunes using aerial photographs.

The species involved in the colonization of the hypersaline substrate of the exposed bed of the drying Dead Sea include Sudanian and Saharo-Arabian elements with tolerance for extremely high salinity. The plant communities formed by these species are relatively recent. The time involved encompasses only the past fifty or sixty years, during which the rapid decline in Dead Sea levels has exposed extensive surfaces of salty terrain. However, within the exposed areas, two main belts have been identified (Aloni, Eshel, and Waisel 1997). The lower belt corresponds to the most salt-tolerant species, which include *Atriplex halimus, A. holocarpa, Salsola baryosma*, and *Anthemis maris-mortui*, among others. The second belt occupies those areas from which salts have already been washed, even by the low precipitation amounts. The vegetation occupying this belt includes typical salt-tolerant species growing elsewhere in the Rift Valley, such as *Zygophyllum dumosum, Tamarix nilotica, Fagonia mollis,* and *Medicago laciniata,* among others. These belts have shifted down in elevation as the process of shoreline recession continues. It is hypothesized that eventually the climax vegetation on the exposed shores will consist of the typical communities dominated by *Zygophyllum dumosum, Acacia* spp., and other species typical of the Sudanian region.

The Oases

The most representative area of oasis vegetation in Jordan is the Azraq wetland in the Azraq Shishan area (fig. 4.1-B). Farther north, the Azraq Druze area used to have a larger wetland but is now mainly dry, except for a few ponds built by the villagers (Nelson 1974).

In the twentieth century, the size of the ponds in the Azraq Shishan area were reduced considerably by pumping, but with the creation of the Azraq Wetland Nature Reserve, the ponds have been restored. The oasis vegetation of this area can be subdivided into three groups on the basis of distance from water sources and salinity (Nelson 1974; Al-Eisawi 1996): freshwater, brackish-water, and grassland vegetation.

The freshwater vegetation is restricted to the main pools, where the water is usually potable because of its low salinity. The most common species are *Typha domingensis, Phragmites australis, Juncus acutus, J. maritimus, Imperata cylindrica, Sonchus maritimus,* and *Inula chrithmoides* (Al-Eisawi 1996).

The brackish-water vegetation grows along canals and bodies of water that contain water only in certain seasons. Water salinity in those localities is considerably higher than in the pools but lower than in the salt flats. In and around the pools, the dominant species along the wadi channels emptying into the Azraq basin are *Ruppia cirrhosa, R. maritima, Zanichellia palustris, Scirpus maritimus,* and the algae *Chara* spp.; around the main seasonal pools, the predominant shrubs are *Tamarix tetragyna, T. amplexicaulis,* and sometimes *T. passerinoides.* Away from the pools, the dominant species are small shrubs such as *Alhagi maurorum, Cressa cretica,* and *Nitraria retusa.*

Grasslands occupy areas subject to frequent water impoundment, which results in rapid accumulation of organic matter. The dominant grass species in this environment is *Phragmites australis,* which is accompanied by small amounts of other annuals such as *Polypogon monspe-*

liensis, Spergularia salina, Juncus bufonius, Sonchus oleraceus, and *Suaeda asphaltica*. In addition, perennial sedges are dominated by *Juncus maritimus* and *J. acutus*.

The year-round green vegetation around ponds continues to attract large numbers of birds and other fauna. The Azraq wetland used to be considerably larger, and the springs that feed it used to be more generous. For millennia, this part of the desert has been a center of attraction for populations of humans and animals, as attested by remains of fauna, flora, and artifacts (Rollefson et al. 1997; Rollefson 2000).

Human Impact on the Drylands

Although most dryland ecosystems are not favorable to most human activities, they have not been spared from human transformation and degradation. Botanical, faunal, and archaeological records provide evidence of landscape transformation by human populations since the Early Holocene. Although transformation occurred in ways different from those that affected the subhumid areas of the highlands, the steppes and deserts of Jordan, like most dryland ecosystems in the world, responded quickly to the effects of human agency.

In the case of Jordan, the Irano-Turanian steppe has been so transformed by the effects of grazing that determining the appearance of pre-agricultural landscapes is difficult. The impact of the wild ancestors of modern domesticated grazing animals was minimal because they occupied specific habitats and predators controlled their population numbers (Clutton-Brock 1981). Over the past nine thousand years, nomadic pastoralism in the Near East has been a strategy of subsistence in areas with low carrying capacity, requiring the use of environments in the different seasons (Wagstaff 1985). Because this system implies seasonal movements to a variety of ecological zones, all natural regions are subject to livestock grazing. Even the inhospitable deserts have not been spared the effects of pastoralism. Areas around oases and springs in more arid areas have been particularly transformed because sources of water are usually focal points for pastoral nomads and sedentary populations.

Sheep and goats are the most common livestock in the Near Eastern drylands. These two closely related species developed different ways of grazing. Sheep graze to a root level, destroying the herbaceous mat to the ground, while goats graze indiscriminately on trees, shrubs, and herbs (Köhler-Rollefson and Rollefson 1990). In the end, goats are the most destructive, since their adverse effects cover larger areas. Because grazing implies the removal of vegetation, soils become unprotected and vulnerable to the effects of erosion by water and wind.

From the ecological point of view, livestock grazing implies the selection of certain species of plants that are preferred by livestock. This means that before the establishment of grazing, the composition of the vegetation in most regions was certainly different than at present. The devastating effects of grazing can be seen almost everywhere throughout the Near East. The differences between nongrazed and grazed landscapes are striking, as can be seen in photographs and satellite imagery in the Sinai and Negev deserts, where the Egyptian-Israeli border marks the limit of grazed and nongrazed sand hills (Kedar 1985). On the Egyptian side of the border, vegetation cover is sparser because Bedouin populations are larger and more nomadic. These differences have led to the onset of "reversed desertification" on the Israeli side, bringing about stabilization of sand dunes in less than twenty years (Tsoar et al. 1995).

Antipastoralism

Because most of the territory of Jordan is exposed to intense livestock grazing, the majority of the remaining herbaceous plants are unpalatable to livestock. The term *unpalatable* refers to plants that are avoided by livestock because they are poisonous, have spines, or contain repelling oils or odors. Collectively, this group of plants is often referred to as antipastoral vegetation.

Vegetation transects in Wadi Rum and Wadi ath-Thamad show, for example, the abundance of antipastoral plants in the drylands of Jordan (fig. 4.3). The Wadi ath-Thamad area is located in a region that receives 150–200 millimeters of precipitation a year on the Irano-Turanian steppe. Vegetation on the uplands on both sides of the wadi is dominated by the unpalatable *Anabasis* scrub and *Artemisia* (fig. 4.3-B). After the first autumnal rains, *Urginea maritima* and various species of *Colchicum* appear, both of which are unpalatable. The bottom of the wadi is dominated by the poisonous *Nerium oleander* and the spiny *Alhagi maurorum* and *Capparis spinosa*. The majority of the plants identified and collected along this transect are listed as poisonous and unpalatable (Abu-Irmaileh 1988). *Peganum harmala*, a highly poisonous plant of the family Zygophyllaceae, is found in almost every region of Jordan. Among other unpalatable plants in this region are the members of the family

Labiatae (mint family), which are often avoided by cattle because of their smell. Most of these plants are not found exclusively among the antipastoral vegetation of the Irano-Turanian steppe but grow in most of the other dryland vegetation regions as well. This suggests how ubiquitous these species have become in the landscape of Jordan.

The Wadi Rum area is located in a hyperarid zone that corresponds to the transition between the Sudanian and Saharo-Arabian vegetation regions of southern Jordan. Despite being an area that receives only about 50 millimeters of rainfall a year, Wadi Rum has an enormous diversity of desert and oasis vegetation (fig. 4.3-A). Most of the elements present in the lowlands and sand dunes are unpalatable plants of the family Chenopodiaceae (e.g., *Hammada salicornica*). *Peganum harmala* is ubiquitous, especially on the alluvial fans. Many of the antipastoral plants indeed are well adapted to extreme aridity as well as other adverse conditions.

The effects of pastoralism on the natural vegetation of the Near East are reflected in most pollen diagrams in the form of a rapid increase of plant taxa associated with grazing. For the Mediterranean zone, these taxa include a variety of herbs and scrub, which in turn can be subdivided into two groups. They include *Plantago lanceolata*, *Rumex acetosella*, and *Urtica* spp., which benefit from nitrates derived from livestock dung, and spiny plants of the rose family, often referred to as *Poterium/Sanguisorba* (Bottema and Woldring 1990). However, the ecology of some of these groups is different in the drylands.

Dryland Cultural Ecology

Culturally, the drylands of Jordan are often associated with nomads and their flocks of sheep and goats, as pastoralism has been the main economic activity of steppe and deserts in this part of the world. However, some of the traditional seasonal movements of populations and livestock in the steppes and deserts have been reduced or modified from their traditional patterns.

The traditional nomadic seasonal patterns still practiced by Bedouins during the early decades of the twentieth century provide a clue to the ecological aspects of husbandry in the past. The ethnographic and historical examples of Bedouin populations in Syria and Jordan provide a picture of how the nomadic populations responded to environmental change (e.g., changes in climate and vegetation) and their complex rela-

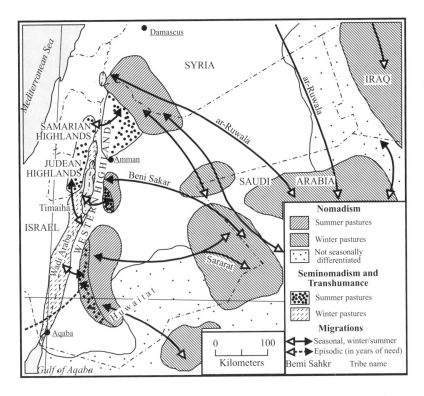

Figure 4.6 Traditional transhumance paths in Jordan and adjacent countries (after Scholz and Schweizer 1992, with modifications by the author).

tionships with settled farmers in the more humid western regions of the country (Lewis 1987).

Before the demarcation of modern political borders, the seasonal migration routes covered long distances in a complex but at times flexible pattern (fig. 4.6). Although these historical seasonal routes correspond to the late Ottoman period and early British Mandate period, they may follow geographic patterns established in earlier times.

Nonetheless, despite the closing of the borders, seasonal migration patterns continue within shorter distances inside each country. Two examples of short-distance nomadism in Jordan are the cases of the Abu Tayeh Bedouins in southern Jordan and the Saba'in on the Madaba Plateau.

The Abu Tayeh Bedouins have practiced vertical nomadism between the limestone uplands of Ras an-Naqb and the lower sandy areas in the Disa basin. This pattern has been identified in archaeological sites dating back to early agricultural times (Henry 1989). Although modern Bedouins are recent Arabian newcomers, they adopted nomadic paths similar to those of groups who lived in the area several centuries earlier.

The Saba'in Bedouins occupy the Madaba Plateau, particularly along Wadi ath-Thamad and its tributaries, an area where I have done extensive geoarchaeological research. Originally from the Negev Desert, the Saba'in came as Palestinian refugees, becoming the guests of one of the local tribes with whom they have blood ties. Their pattern is also vertical, involving movements to the valleys and canyons in the winter and the plateau during the summer. But this pattern was modified as they began to use small-scale irrigation schemes and fodder bought from irrigated farms. In the same fields where they cultivate rain-fed barley in the winter, they have begun to farm watermelons, cantaloupes, and a variety of vegetables using gas-powered pumps to extract water from wadi beds during the summer.

How Much Farming Do Nomadic and Seminomadic Groups Practice?

This is a question that has prevailed in the literature on rural subsistence in the drylands of the Near East. Remains of cultivated varieties of cereals recovered from sites apparently occupied by pastoral groups raise the question of whether these groups practiced some form of farming or obtained cereals from settled farmers in humid areas. This is the case of Ain Abu-Nukhaylah, a Pre-Pottery Neolithic site located in an area with insufficient rainfall (Henry et al. 2003).

The information that archaeological records provide suggests that some of the so-called nomad groups practiced a mixed economy based on pastoralism and farming. These groups did not have permanent settlements such as the towns and villages of sedentary farmers, but their economy was mixed, complex, and variable, depending on the rain outcome of a particular year.

Other activities beyond husbandry were also practiced by nomads in the past. Research in the Negev has shown the economic complexity of nomadic groups in prehistory and history as more nomadic sites have been excavated. This research shows, for example, the essential role of nomadism in the development of towns and foreign trade in the Bronze Age (Rosen 2003) and the functioning of regional Byzantine and Early Islamic economies (Rosen and Avni 1993), which contradicts the common assumption that nomadism has been merely focused on the production of milk and meat.

Studies of historic and modern Bedouins by Norman Lewis (1987) and Kenneth Russell (1995) explain how nomad groups practice complex mixed economies, contradicting the popular notion that pastoral nomads merely wander in search of water and grazing grounds. Lewis (1987) reported that in the Syrian and Jordanian deserts, people of each tribe were intimately familiar with their *dīrāh*, or the area where they normally moved with their flocks or herds. Therefore, there existed a profound knowledge of where areas for potential farming were located. Bedouin farming was not a spontaneous activity, since it involved a certain degree of planning and decision making. Grazing and farming were activities that varied depending on group decisions, intelligence gathering, and the status of intertribal alliances (Lewis 1987).

The traditional dīrāh system described by Lewis (1987) has now been modified because of the use of motorized vehicles that facilitate mobility and, most important, the use of water pumps, which procure water at one place. In addition, mobility has been reduced because of boundary closing and government policies that encourage permanent settlement.

Despite the modern changes in the nomadic way of life, the typical *khaymah* (tent) is still one of the most common cultural features of the Jordanian rural landscape. Tents can also be seen in towns and sometimes even in empty lots within the Amman metropolitan area. In some modern Bedouin settlements, it is common to see tents next to concrete or cinder-block houses. Family activities such as tea drinking, gatherings, and cooking often take place in the tent. The shaded and cooler space created by the tent provides an advantage over the modern concrete houses.

Camels, the main vehicles for nomads since the Iron Age, are now being replaced by pickup trucks. But camels remain an important part of Bedouin life; they are still used for transportation in difficult terrain. Additionally, camels are still an asset for most Bedouins, who very often include them as part of a dowry. The frequent road signs warning drivers of wandering camels are an indication that camels remain a part of the Jordanian landscape.

Although the deserts of Jordan are the least populated parts of the country, settled areas within them are on the rise. The result of encouraged sedentary life and the rapid increase of the country's population have expanded the number of settlements in the desert. Two areas of the deserts with recent increase in settlement are the Eastern Badia and Al-Jafr basin. Of the two, the Eastern Badia has seen more significant development in the past two decades.

The Eastern Badia includes the strip of Jordanian territory that extends from the Azraq basin to the Iraqi border (fig. 4.1-A). The availability of water resources, the pipelines that crisscross the region, and the ever busier Amman–Baghdad road are the main factors that have encouraged settlement growth in this region. Recent economic and political changes in Jordan are rapidly transforming the traditional systems of the communities in the Eastern Badia (Millington, Al-Hussein, and Dutton 1999). However, according to Alan Rowe (1999) such changes have not affected the ways in which the landscape is exploited by the Bedouins, particularly because a mixed economy of farming and pastoralism is still fundamental. However, the new feedlot system involving alfalfa and other nontraditional crops for feed are becoming more common in Jordan, thus eliminating a millennial system of nomadic pastoralism. This represents an example of how traditional groups simultaneously adapt to modern life without abandoning traditional subsistence methods.

Ancient Settlements in the Drylands

A variety of ancient settlement types can be found in the Jordanian drylands. These include campsites, larger seasonal settlements, forts, permanent villages, and even towns. One of the best examples of the variety of settlements is in the Eastern Badia, where numerous surveys have provided evidence of cultural continuity and change over millennia. During the Neolithic, the area was characterized by two categories of sites: permanent stations and seasonal camps (Betts 1998). In addition, the rocky terrain allowed for several hunting strategies, the most typical being the hunting enclosures known as *kites* (Betts 1998).

In later periods, settlements diversified based on their economic specialization, location, and available resources. The practice of floodwater irrigation allowed an increase in agricultural production and subsequently larger settlements. Economic, political, and social factors were also important, as the Eastern Badia included some of the *limes* (forts) guarding the Roman eastern frontier (Kennedy and Riley 1990). Changes in trade routes during the Byzantine and Early Islamic periods also influenced settlement patterns in the area. Located at a crossroads between Syria, Iraq, the Arabian Peninsula, and Palestine, the area became more attractive for settlement, especially around springs and oases. The castles built during the Islamic period provide evidence of the relative prosperity and economic importance of this region.

Archaeological records reveal that the Jordanian steppes and deserts were not always devoted to pastoralism. Several locations in the drylands of Jordan display evidence of early irrigated agriculture and urbanism. One example is the flood-irrigation system built circa 3000 BC along Wadi Rajil in the Black Desert (al-Harra) of northeastern Jordan (fig. 4.1-A). This system consists of a series of diversion walls and dams associated with the Chalcolithic–Early Bronze settlement of Jawa, studied by Svend Helms (1981) and later excavated by Alison Betts (1991). This system's function is similar to flood-irrigation systems in the Negev (Evenari, Shanan, and Tadmor 1982), although the installations associated with Jawa predate those of the Negev.

Wadi Faynan, located in the hot and dry environment of Wadi Araba, is another region that has received a great deal of attention by archaeologists because of its long-term importance as a center for copper extraction and smelting. The amount of population sustained by the region necessitated that food be produced to feed the residents. Therefore, the region also became an important agricultural center. Water diversion installations and ancient agricultural fields have been reported on alluvial fans of the Wadi Faynan region (Barker 2000; Barker et al. 1997, 1998, 1999). Although the system was used through several periods, irrigation may have begun as early as the third millennium BC.

The best example of a culture intensely involved with water management structures is that of the Nabataeans, a civilization that excelled in agricultural technology, water management, urbanism, and trade in the second and first centuries BC. Their independence ended with their inclusion into the Roman Empire and its successor the Byzantine Empire, during which the Nabataeans remained as masters of water in the desert.

The Nabataean Kingdom occupied areas of present-day Jordan, Saudi Arabia, Israel, and Egypt (the Sinai Peninsula only), a territory consisting mainly of areas that today receive less than 200 millimeters of annual rainfall (fig. 4.7). The Nabataeans' success with managing water seems to have been a result of their engineering skills within an environment that

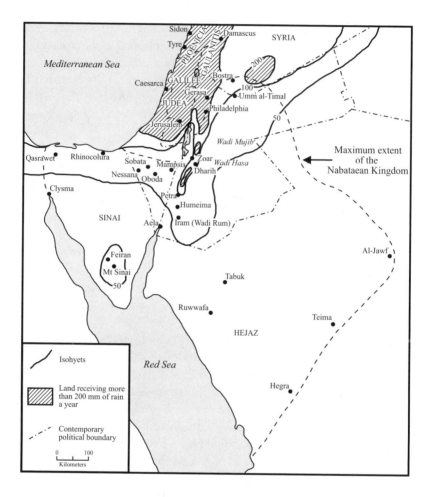

Figure 4.7 The Nabataean cities and the maximum extent of the Nabataean Empire in the drylands of the Near East. Limits of the Nabataean Empire are based on Taylor 2002. Isohyets show current average annual precipitation.

was probably more humid than today's. Paleoclimatic reconstructions point to a more humid period during the second and first centuries BC (Bruins 1994). Nonetheless, exactly how much more annual precipitation this area received at that time is not clear.

Most archaeological studies focus on architecture and other cultural aspects of the Nabataean culture. Therefore, little research has been conducted on the environmental context of this civilization. A few studies, however, have addressed hydrologic aspects linked to urbanization and agriculture. The best-known case of architecture and water-management engineering is that of Petra, where today one can still see canals carved in the sandstone, as well as numerous remains of dams.

Another prominent case of Nabataean irrigation is that of Humeima (ancient Havarra), where a team from the University of Victoria, British Columbia, has led a research project on water management and irrigation systems. Humeima is located in a valley at the northwestern edge of the sandstone country of southern Jordan. Like most Nabataean cities, Havarra was located along the main trade routes. Although the town was originally founded around a small spring, city dwellers subsequently built a water supply system highly integrated with the settlement design (Oleson 1997). The water system in Havarra was designed to use topography to direct water to two public reservoirs and various private and domestic cisterns. This example of effective use of water illustrates how ancient engineering knowledge contributed to the maintenance of high population levels in the desert.

5

Paleoecological and Geoarchaeological Records
Current Status and Prospects

Despite the rapid growth of research in the areas of geoarchaeology and Quaternary paleoecology, paleoenvironmental records for Jordan are still scanty. Most of the extant records are focused only on small areas, usually around archaeological sites. In part this is because archaeological research, the main broker for paleoenvironmental research in Jordan, has focused on study areas within the limits of archaeological sites and surveys. Despite their scarcity, paleoenvironmental records in Jordan have already provided a strong basis for interpretations and opened the path for future research.

Global, Regional, and Local Records

Local and regional paleoclimatic sequences for the Quaternary period are often correlated with oxygen isotope data registered in deep-sea and ice cores. The major fluctuations of stable oxygen isotope ratios ($\delta^{18}O$) from fossil organisms deposited in deep sediments of the world's oceans are directly related to global climatic changes. On the basis of temperature fluctuations, Cesare Emiliani (1972) assigned odd numbers to warmer periods (interglacials and interstadials) and even numbers to colder periods (glacials). These major divisions are known as marine isotopic stages (MIS), which for the past 140,000 years comprise six stages (fig. 5.1). Later on, Nicholas Shackleton and Neil Opdyke (1973) defined five subdivisions within MIS 5.

Although several cores have been taken from the South Pacific, Caribbean, and North Atlantic, they all present similar fluctuations (Williams et al. 1998). Thus, the curve of ocean paleotemperatures from the North Atlantic (Core V23-82) (Sancetta, Imbrie, and Kipp 1973) has been correlated with a core from Barbados (Bradley 1999).

Oxygen isotope records are also obtained from cores in ice sheets remaining in Greenland, the Canadian Arctic, and Antarctica, as well as from ice caps on various mountains in the world. These ice cores can be

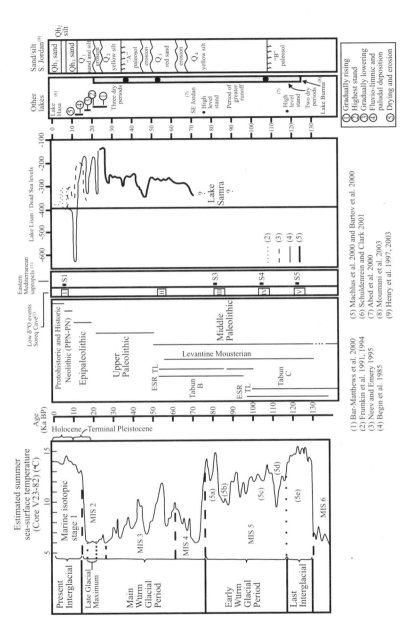

Figure 5.1 Estimated sea-surface temperatures in the North Atlantic (based on Bradley 1999), lake-level changes in the Dead Sea basin, and other lacustrine sequences and eolian deposits in relation to the broad cultural chronology of the southern Levant for the Late Quaternary (based on Bar-Yosef 1998 and Henry 1997b).

as long as 2 kilometers, comprising thousands of years. As with the deep-sea cores, $\delta^{18}O$ values for the ice cores are converted into temperatures. Global temperature changes recorded in ice cores are often correlated with climatic fluctuations elsewhere (Alley 2000). For example, dry episodes in the Near East, such as the Younger Dryas, have been linked to low $\delta^{18}O$ values in the Greenland ice cores (Weiss 2000).

Although paleoclimatic data derived from deep-ocean and ice cores reflect global changes, they are often used as a reference for explaining regional and local changes recorded in terrestrial environments anywhere in the world. Thus, in the southern Levant, some correlations can be drawn with lake-level changes of Lake Lisan and the Dead Sea (fig. 5.1). However, local changes sometimes differ from the global temperature change. Therefore, caution should be taken when extrapolating changes to the marine isotope stages curve.

Although the Levant was not glaciated, paleotemperatures played a role in changing atmospheric moisture. As a general pattern, cold events meant dryness, while the opposite can be said for warm stages and events. However, local variations of this pattern have been observed. For example, the Last Glacial Maximum (21–16 ka BP) was dry for most areas in the Mediterranean region. However, reports of moist short-lived events occurred in various desert areas of the southern Levant, particularly toward the end of the Last Glacial Maximum (Henry 1997b). Unfortunately, the scarcity of paleoclimatic records in Jordan forces scholars to draw most of the proxies for paleoclimatic reconstruction from records elsewhere. Potential sources of paleoclimatic information exist everywhere around the country, although poor preservation of records and difficulties in dating certain environments present serious challenges to this type of research.

Pollen Records

Fossil pollen sequences in the Near East have been obtained chiefly from lake deposits in Turkey, Iran, northwestern Syria, and northern Israel. Because of the lack of long pollen records in Jordan, the pollen sequences from lakes in the Rift Valley and the Golan Heights are often used as a proxy for regional vegetation reconstruction in western Jordan (fig. 5.2).

As part of the *Tübinger Atlas des Vorderen Oriens*, a series of paleo-vegetation maps (Van Zeist and Bottema 1991) were published for the

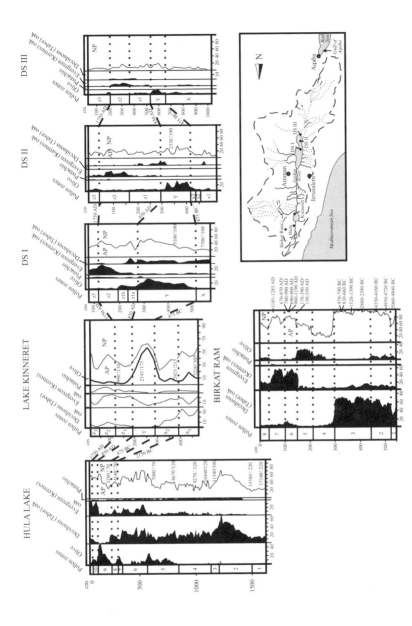

Figure 5.2 Pollen diagram summaries from lakes in the Jordan–Dead Sea Rift Valley. (Lake Kinneret is after Baruch 1990; DS I, DS II, and DS III are from Baruch 1993; Hula Lake is after Baruch and Bottema 1999; and Birkat Ram is after Schwab et al. 2004).

southern Levant (Israel, Palestinian territories, and western Jordan) based on pollen data from Lake Hula (Tsukada in Van Zeist and Bottema 1991), Birkat Ram (Weinstein 1976), and Lake Kinneret (Baruch 1990). Since their publication, these maps have been the main reference for regional vegetation reconstruction during the Terminal Pleistocene and the Holocene in Palestine and western Jordan.

The main information sources that update our knowledge of southern Levantine vegetation are provided by more recent pollen data (Baruch and Bottema 1991, 1999; Baruch 1993; Heim et al. 1997; Schwab et al. 2004). Additionally, a new reconsideration of the dates in a core (Baruch and Bottema 1999) in Lake Hula has been proposed (Meadows 2005). Unlike the previous pollen diagrams from the region, the new Birkat Ram core and the Dead Sea DS7–1SC core provide relatively high resolution for vegetation changes that occurred during the past two millennia, in particular during the most recent centuries.

Within the territory of Jordan, as in most of the arid and semiarid Middle East, pollen-derived vegetation reconstructions are basically absent because of the scarcity of lacustrine deposits bearing fossil pollen grains. Attempts have been made to study pollen from playa-lake sediments of the Central Plateau, but such deposits have shown poor preservation and concentrations too low for paleovegetation reconstruction (Caroline Davies and Stephen Hall, personal communication). The difficulties in obtaining pollen from sedimentary deposits in arid regions are attributable to poor pollen production of most local plants and poor pollen preservation as a result of the extreme weather phenomena typical of these regions. But despite these problems, palynologists have obtained pollen records from deposits previously believed to be unsuitable for pollen studies (Horowitz 1992).

Pollen and spores for landscape reconstruction in Jordan have been extracted from a variety of deposits, including cultural (archaeological), alluvial, floodplain wetland, palustrine, and zoogenic (fig. 5.3; table 5.1). Nonetheless, these studies have produced mixed results, creating doubt as to whether pollen data from nonlacustrine deposits are reliable for the reconstruction of paleovegetation.

Archaeological deposits are among the most frequent sources of pollen for vegetation reconstruction, despite the disagreement among palynologists with regard to uniform preservation and significance of fossil pollen assemblages from cultural deposits for reconstructing paleovegetation. A series of examples in Levantine sites shows that archaeological

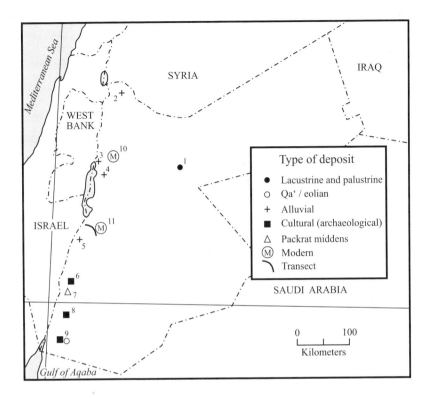

Figure 5.3 Location of Late Quaternary palynological studies in Jordan. Refer to table 5.1 for details.

deposits, as with most terrestrial sediments, are not as rich in pollen as marsh and lake sediments are, but they still yield enough counts for a diagram (Leroi-Ghouran 1982). However, criticism of pollen assemblages from archaeological deposits points to overrepresentation of certain groups of plant taxa resulting from selective deterioration and transport qualities (Bottema 1975). For example, pollen grains of Asteraceae Liguliflorae (i.e., the dandelion group) are often found in large amounts in archaeological deposits not because of their local abundance but because of their great resistance to deterioration (Bottema 1975).

Nevertheless, considerable amounts of pollen have been obtained from prehistoric deposits in rockshelters in the arid areas of south Jordan (Emery-Barbier 1995). Similarly, pollen was successfully obtained from archaeological deposits from the Epipaleolithic and Neolithic site

Table 5.1 Published Quaternary pollen studies in Jordan and the Rift Valley

Locality	Number on map	Type of deposit/study	Time frame or cultural periods	Publication
Ain al-Assad	1	Palustrine/spring	Various periods within the past 100 ka BP	Kelso and Rollefson 1989
Wadi Shallalah	2	Wetland/alluvial	6–0 ka BP	Fig. 5.3
Wadi Kafrein	3	Alluvial	Medieval	Vita-Finzi and Dimbleby 1971
Wadi al-Wala	4	Alluvial	Middle Holocene/ Medieval, Modern	Fig. 5.3
Wadi Faynan	5	Alluvial	Early to Middle Holocene	Barker et al. 1999, Hunt et al. 2004
Beidha	6	Cultural deposits	Natufian	Fish 1989
Petra/Wadi Musa	7	Hyrax middens	Historic	Fall 1990, Fall et al. 1990
Jabal Qalkha	8	Cultural deposits	Middle and Upper Paleolithic, Epipaleolithic	Emery-Barbier 1995
Ain Abu-Nukhaylah	9	Cultural and qaᶜ deposits	Pre-Pottery Neolithic B	Scott-Cummings 2001
Amman area	10	Traps	Modern	Al-Eisawi and Dajani 1987, 1988
Southern Highlands to Wadi Araba	11	Surface samples along altitudinal transect	Modern	Davies and Fall 2001

Note: The abbreviation "ka BP" denotes the number of thousands of years before the present.

of Beidha, located north of Petra (Fish 1989). Fossil pollen was also obtained from lake, marsh, and alluvial deposits at the prehistoric site of Ain al-Assad in the Azraq basin (Kelso and Rollefson 1989).

Alluvial sediments often yield fossil pollen that is useful in paleoecological reconstructions. Despite the obvious problems with reworking and selection by water currents, alluvial pollen has proved effective if interpreted properly. Claudio Vita-Finzi and Geoffrey William Dimbleby (1971) provided one of the earliest attempts to interpret alluvial pollen spectra in Jordan. Studies such as those from Ain Al-Assad (Kelso and Rollefson 1989) and Wadi Faynan (Barker et al. 1999) have shown that even in arid areas, concentration of pollen in alluvial deposits can be sufficient for reconstructing paleovegetation. The success of using alluvial fossil pollen for paleoecological studies lies in finding the right alluvial deposit and using the appropriate method for pollen extraction.

I carried out a series of procedures to obtain pollen from alluvial deposits in Jordan. Preliminary data showed that the best preservation and concentration of pollen occurs in sediments deposited by relatively steady and calm waters (low energy of deposition), usually in low-gradient floodplains or behind natural or artificial dams. The disadvantage, however, is that alluvial stratigraphic sequences often present chronological gaps. Consequently, alluvial sequences bearing pollen encompass only relatively short periods of time. However, if such sequences are properly dated, the pollen assemblages obtained from them can provide useful information on vegetation and land use for certain periods comprising several centuries and even millennia. Two of the alluvial sequences that I worked on in Jordan were from Wadi al-Wala and Wadi Shallalah (fig. 5.4). Mid-Holocene deposits in the two sequences represented by the Iskanderite Alluvium in Wadi al-Wala and Units I and II in Wadi Shallalah seem to correlate with the regional picture of tree pollen recorded in the pollen records from Hula and Birkat Ram, in which *Olea* pollen appears to be abundant during the time comprising the Early Bronze Age (cf. fig. 5.2). Additionally, deciduous oak seems to dominate over evergreen oak in both the local (Wadi al-Wala and Wadi Shallalah) and regional records.

Although recent alluvial deposits in Jordan may contain considerable amounts of moderately preserved pollen, alluvial deposits of Pleistocene age have very poor preservation and concentration or, in most cases, no pollen at all. My suspicion is that intense pedogenesis involving clay formation during the wet phases of the Terminal Pleistocene (15–12 ka

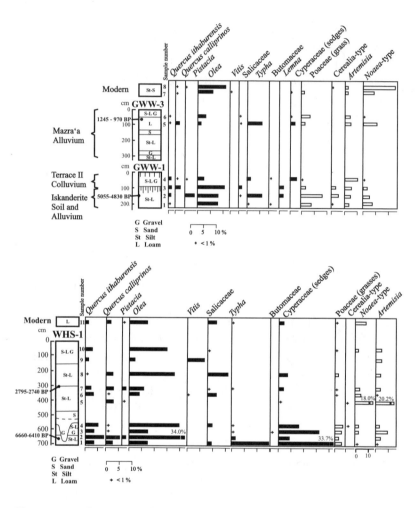

Figure 5.4 Pollen records from Middle and Late Holocene alluvial deposits in Wadi al-Wala (*top*) and Wadi Shallalah (*bottom*).

BP) led to pollen grain deterioration. However, I have found excellent pollen preservation from spring and fluvio-lacustrine deposits around the Azraq wetland.

Quaternary travertine and tufa are among those deposits containing fossil pollen that have been studied in the drylands of the Levant (Weinstein-Evron 1987). The abundant travertine and tufa deposits in Jordan thus may also provide acceptable pollen counts. Some localities along streams dissecting the Irbid Plateau have massive sequences of travertine that could provide a continuous record. One such sequence is in Wadi Abu-Said (El-Radaideh 1993).

Paleovegetation reconstructions based on pollen extracted from hyrax middens have also proved very useful, although few of them have been tested. The Syrian rock hyrax (*Procavia capensis* subsp. *syriaca*) constructs middens composed of plant fragments, fecal pellets, and faunal parts, bound together by urine (Fall 1990; Fall, Lindquist, and Falconer 1990). Besides the numerous botanical remains collected by the hyrax, pollen grains adhere to the sticky surfaces of the midden, which builds up over time. The paleobotanical information obtained from hyrax middens is twofold: pollen provides a broad idea of the regional vegetation, while plant macroremains provide information on plants that existed within a short distance of the midden. Hyrax middens collected from the Petra region provided data for the reconstruction of local vegetation, which in turn presents parallels with lacustrine pollen cores in the Dead Sea and Lake Kinneret (Falconer and Fall 1995).

A customary way to interpret fossil pollen assemblages is to compare them with modern assemblages in various ecologically different areas. Studies on modern pollen spectra have the purpose of linking modern pollen rain with extant vegetation, which in turn serves as a reference for interpreting fossil pollen assemblages. In the Levant, such studies have pointed to problems of pollen production, preservation, and most important, foreign pollen influx in the local assemblages. The most notable of such studies in the Levant are Horowitz 1969, Horowitz, Weinstein, and Ganor 1975, and Weinstein 1979 in Israel; Bottema and Barkoudah 1979 in Syria; and Davies and Fall 2001 in southern Jordan. The study of monthly pollen rain (Al-Eisawi and Dajani 1987, 1988) also has contributed to our understanding of the seasonal production and transport of pollen in Jordan.

Modern pollen studies help with not only establishing modern analogs of vegetation but also understanding problems with under- and overrepresentation of certain pollen grains in fossil assemblages. Chenopods are a frequently overrepresented taxonomic group in pollen assemblages of the Levant. Modern pollen rain studies in Israel show that such large amounts of chenopod pollen come with winds originating in the deserts (Weinstein 1979). Therefore, the large amount of chenopods in nondesert areas of the Levant does not indicate local production.

Another issue regarding overrepresentation of certain taxa in fossil pollen diagrams pertains to olive pollen. Pollen diagrams from Hula Lake (Baruch and Bottema 1999) and the Ghab Valley (Yasuda, Kitagawa, and Nakagawa 2000) show *Olea* pollen during the Terminal Pleistocene. This is not uncommon in the Levant, where olive trees grew wild in the Mediterranean forest. However, relatively high amounts of olive pollen have been reported in preagricultural deposits in the arid lands of southern Jordan (Emery-Barbier 1995; Scott-Cummings 2001). Most probably this occurrence implies long-distance transportation. The amounts of *Olea* pollen in modern samples are usually high, even in those areas with a low density of olive cultivation (Bottema 1991; Davies and Fall 2001). Olive trees produce anemophilous pollen (i.e., wind-aided pollination), which is often produced in abundance (Bottema 1991). Therefore, despite the widespread cultivation of olive during the Holocene, *Olea* pollen could possibly be overrepresented because of its high production and dispersion capabilities. In view of this issue, my research is currently focusing on assessing *Olea* in fossil pollen assemblages using modern analogues.

Evidently, more research on modern pollen assemblages and spectra as a reflection of modern vegetation is needed for interpretation of fossil pollen assemblages in Jordan. Because of the preservation problem, studies on pollen taphonomy to identify patterns of pollen preservation are needed, especially on samples from nontraditional deposits such as tufas, alluvium, hyrax middens, and archaeological deposits.

Plant Macroremains

Archaeobotanical research, or the study of plant macroremains from archaeological deposits, presents some advantages over pollen analysis. Seeds, dried fruits, and other plant parts provide more detailed information for the identification of lower taxonomic levels, which in most cases is difficult or impossible using pollen (Miller 1997). Seeds and other remains also provide better morphological clues to differentiate cultivated and wild varieties, which are often unclear in pollen assemblages. While pollen provides a general reconstruction of plants growing in the area around the site and in its immediate regional context, plant macroremains provide information on available plants and plants brought by humans and animals into a site. Plant macroremains are abundant in archaeological deposits, especially as charred material, which facilitates radiocarbon dating. This advantage allows establishing relatively tight chronological control, which is usually not the case with pollen sequences (Miller 1997).

The archaeobotanical record in Jordan is already vast because the majority of sites of most archaeological periods have yielded plant macroremains (see Neef 1997). As in other regions of the Near East, plant macroremains recovered from Transjordanian Epipaleolithic and Neolithic sites provide valuable information on the development of plant domestication. Plant macroremains from Neolithic sites and later periods provide information on the local agricultural economies and trade with other regions.

Assemblages of seeds, wood, and other plant remains also offer valuable information for reconstructing the degradation of natural vegetation around the studied settlements. Seeds of typical weeds associated with farming and grazing are an indirect means by which to reconstruct the sequence of landscape transformation by early agricultural settlers (Hillman 1996; Miller 1997). For example, the Pre-Pottery Neolithic assemblages of burned wood at the site of Ain Ghazal show the incremental use of oak (*Quercus ithaburensis*) during the Pre-Pottery Neolithic phases (Köhler-Rollefson and Rollefson 1990). The site is located in a treeless area dominated by garrigue scrub and an assemblage of plants typical of the Irano-Turanian steppe.

The study of ancient charcoal as fuel belongs to a subfield in archaeobotany known as anthracology, whose focus is mainly on the reconstruction of forests and woodlands. The advantage of this research field is that some woods can be identified to species level, which makes anthracological studies complementary to pollen analysis (Baruch 1999).

Phytoliths

Phytoliths are microscopic bodies of amorphous silica (opal) deposited in cell walls of stems, leaves, roots, and inflorescences of plants. When plants die and decompose into the soil, opal phytoliths are often preserved in sediments. Phytoliths are resistant to the types of degradation processes that destroy pollen and macroremains. Therefore, they tend to remain in deposits where poor organic preservation exists, thus presenting an advantage over studies of pollen and plant macroremains. But where pollen and plant macroremains are preserved, phytoliths may add supplementary and comparative information. One of the main contributions of phytolith research in Near Eastern sites is the identification of wild and cultivated grasses and their paleoenvironmental conditions, including seasonality and the identification of ancient irrigation (Rosen and Wiener 1994). Phytolith analysis has aided in reconstructing the presence of different subfamilies of grasses in paleoenvironments, which is not possible through pollen analysis.

Despite their potential in reconstructing Near Eastern environments, phytolith studies have hardly been studied in Jordan. However, three studies worthy of mention are Buehrle 1992 from late prehistoric cultural deposits in Wadi Ziqlab, and Rosen 1995 and Rosen 2003 in prehistoric cave deposits in Jebel Qalkha.

Tree-Ring Research

Dendrochronology, or the study of tree rings, is a technique used for both dating and reconstructing past climatic conditions. Tree-ring research provides year-to-year paleoclimatic data, which is not possible using other paleobotanical techniques. Although popular in many regions of the world, tree-ring studies are scarce in the Middle East because of the low number of species with dendrochronological potential and their low presence in archaeological contexts (Liphschitz 1986).

One of the pioneer studies in Levantine dendrochronology is the study of Phoenician juniper (*Juniperus phoenica*) tree rings in the southern Negev and Sinai by Nili Liphschitz (1986). Ramzi Touchan and Malcolm Hughes (1999) carried out dendrochronological studies for paleoclimatic reconstruction for the past three centuries using tree rings of *Pinus halepensis, Quercus aegilops* (a variety of *Q. ithaburensis*), and *Juniperus phoenica* in various parts of the western highlands of Jordan. The results of this research were correlated with recent meteorological and water resource data, which in principle shows the relatively high accuracy and resolution that this method of paleoclimatic reconstruction represents. However, the main disadvantage of dendrochronology for paleoclimatic reconstruction is that tree-ring chronologies in the Levant are limited to recent historical periods.

Faunal Remains

Remains of wild animals can be associated with specific vegetation biomes and climates in the past. In some cases, faunal remains are good indicators of paleovegetation in areas where palynological and paleobotanical remains have not been recovered. For example, prehistoric assemblages of faunal remains are associated with Lower Paleolithic material from Qaᶜ Azraq and other paleolacustrine basins in the Syrian desert. In Ain Soda, near Azraq Shishan (see fig. 4.1-B), faunal remains associated with hand axes and other late Lower Paleolithic materials include elephant, rhinoceros, and other animals typical of the African savanna (Rollefson et al. 1997). This evidence suggests that during interglacial and interstadial periods of the Middle Pleistocene, this desert region was better endowed with water and the climate was somewhat more tropical. The faunal findings have stirred my interest in linking fauna with flora. I am starting an investigation on paleovegetation in search of tropical savanna flora using pollen and phytolith analysis from the same deposits in which the faunal remains were recovered.

In addition to paleoclimates and paleovegetation, faunal remains provide information on subsistence, diet, and a number of cultural ecological patterns of past societies. Hunting patterns are among the main cultural ecological pieces of evidence provided by faunal assemblages. Terminal Pleistocene sites around Lake Hasa have provided sizable amounts of faunal remains used to reconstruct hunting patterns and wildlife landscapes around the lake (Coinman 1998; Olszewski and Coinman 2002). This information also shows how the disappearance of the lake at the beginning of the Holocene changed the fauna of the region.

Numerous examples of studies on faunal assemblages parallel the information provided by pollen and archaeobotanical remains with respect to human-induced environmental degradation. For example, faunal remains recovered from the Neolithic site of Ain Ghazal include species typical of steppes, woodlands, riverine forests, and standing water (Köhler-Rollefson and Rollefson 1990). The landscape in this region today is far from what the faunal remains show. Therefore, one can gain an idea of how different the landscape was during the occupation of the site. Faunal remains from this site also produced information on the dramatic reduction of biodiversity between the Pre-Pottery Neolithic B (7250–6000 BC) and the Pre-Pottery Neolithic C (6000–5500 BC) (Köhler-Rollefson and Rollefson 1990; Köhler-Rollefson, Quintero, and Rollefson 1993).

The faunal records also show that the reduction of wild species is inversely proportional to the increase of domesticated animals. In the same example of Ain Ghazal during the Middle Pre-Pottery Neolithic B period (7250–6500 BC), one-half of the remains correspond to domesticated goat, which indicates that a meat diet was highly dependent on this animal (Köhler-Rollefson and Rollefson 1990; Von den Driesch and Wodtke 1997). This approach in turn results in valuable information for reconstructing the process of animal domestication and its relation to mobility and farming patterns.

The preservation of animal remains is affected by soil conditions and other environmental factors, as is the case of plant micro- and macro-remains. This is particularly a problem for bones of certain groups such as birds and fish, which do not preserve well in the harsh environment of the Near East (Gilbert 1995). Despite these problems, numerous sites in Jordan have provided large amounts of bones. For prehistoric periods, a list of studies and references to them are presented in Garrard and Gebel 1988, Gebel, Kafafi, and Rollefson 1997, and Henry 1998a.

Although the majority of faunal remains research in Jordan focuses on prehistoric periods, numerous studies have also obtained relevant information from protohistoric and historic sites. Among the various examples of faunal remains from the Early Bronze Age are those from Numeira and Bab edh-Dhra' (Finnegan 1981) and Tell el-Hayyat (Metzger 1984). Some significant examples of faunal remains from historical sites include the study at the Lejjun Barracks (Toplyn 1987), the Faris site on the Karak Plateau (Johns et al. 1989), and the Madaba plains (LaBianca and Von den Driesch 1995).

Assemblages of faunal remains through historical periods in Jordan show an interesting sequence related to cultural preferences, as is the case of the decline in pork consumption during the beginning of the Islamic period (LaBianca 1990). The comparison between the amounts of cattle and ovi-caprids is also a way to assess the role of husbandry and the quality of pastures in historical times (Johns et al. 1989). These examples show how faunal remains offer insights into natural and cultural patterns of landscape change.

Soils and Paleosols

Despite being a useful tool in paleoclimatic reconstruction, soil and paleosol studies have hardly been incorporated into geoarchaeological and Quaternary research projects in Jordan. However, studies already implemented (Schuldenrein and Clark 1994; Henry 1997a; Niemi and Smith 1999; Cordova 2000; Maher and Banning 2001; Moumani, Alexander, and Bateman 2003; Cordova et al. 2005) show the tremendous potential of soil profile studies for landscape reconstruction in Jordan.

Some soil studies have set the basis for future geoarchaeological research. A climatic distinction of soils in Jordan established in Moorman 1959 has served as a basis for subsequent studies. Still other studies (Lacelle 1986; Khresat, Rawajfih, and Mohamad 1998; Cordova 2000; Khresat 2001; Cordova et al. 2005) have recognized the recurrence of argillic and calcic horizons in the soils of the western Jordanian plateaus as a potential source of information for reconstructing paleoclimates and assessing prehistoric soil erosion (Cordova 2000; Krhesat 2001). In the Jebel Qalkha area of southern Jordan, carbonate morphology and other pedogenic features in eolian deposits have been useful for relative dating and paleoclimatic change (Henry 1997a). The potential of secondary carbonate morphology in Late Quaternary soil profiles in Jordan is immense, since some of the calcic horizons are associated with lithic materials (Cordova et al. 2005). The combined use of lithic chronology and luminescence dating could be applied to establishing a pedostratigraphic chronology that could eventually be used for regional correlation and relative dating. An analog to this potential chronology for Jordan is the one developed for the American Southwest based on a series of stages of secondary carbonate development (i.e., Birkeland 1999).

From the point of view of cultural ecology, the study of soils and paleosols could contribute greatly to the understanding of soil as a resource for agricultural groups and the potential of specific soils for rainfed and irrigated farming. Additionally, an understanding of soil profiles could help researchers better understand processes of land degradation such as soil erosion and salinization.

Alluvial Fills

Claudio Vita-Finzi established the first regional chronology of stream aggradation and incision for Jordan; this chronology was based on his pan-Mediterranean model, which recognized a prehistoric Older Fill and a historic Younger Fill (Vita-Finzi 1964, 1966, 1969). Accordingly, the Older Fill occupies a higher elevation over the present stream channel, and it often contains remains of Late Pleistocene archaeological material. The Younger Fill is located at a lower elevation and contains Roman pottery fragments. Criticism of this model arose in several regions of the Mediterranean where numerical dates were obtained from alluvial deposits, showing the presence of more than just two fills (Grove 1997).

Vita-Finzi's model for Jordan later evolved into a four-fill model (Copeland and Vita-Finzi 1978), but because of inadequate dating, this model remained too simple to be used as a regional alluvial chronological sequence for Jordan. Despite its simplicity, Vita-Finzi's model stirred interest in studying the association between prehistoric settlement and cut-fill cycles as evidence for landscape change. Thus, during the past two decades, several geoarchaeological projects have produced abundant information on cycles of alluvial accumulation and erosion in the Late Quaternary. Most geoarchaeological alluvial studies are concentrated in valleys of the Western Highlands and the Rift Valley. These include the Wadi al-Karak and other drainages emptying into the Dead Sea (Donahue 1981, 1984, 1985; Donahue, Peer, and Schaub 1997), Wadi al-Hasa (Donahue 1988; Schuldenrein and Clark 1994, 2001; Neeley et al. 1998; Olszewski and Hill 1997), Wadi Zarqaʾ (Baubron et al. 1985; Besançon and Hours 1985), Wadi Hammeh (Macumber and Head 1991), various eastern tributaries of the Jordan River (Mabry 1992), Wadi Ziqlab (Banning et al. 1992; Field and Banning 1998; Maher and Banning 2001), Wadi al-Ghurab at Beidha (Field 1989), and Wadi ath-Thamad and Wadi al-Wala (Cordova 1999a, 1999b, 2000).

In the steppe and desert regions of eastern Jordan, various sediment chronologies and terraces have been reported for Wadi Jilat (Garrard et al. 1988) and Wadis Enoqiyah, Rattam, and Uwaynid (Besançon and Sanlaville 1988; Copeland and Hours 1988; Besançon, Geyer, and Sanlaville 1989). In the southern deserts, new geoarchaeological studies of alluvial deposits include Wadi Judayd (Hassan 1995), the Jebel Qalkha drainages (Henry et al. 1996; Henry 1997a), Wadi Faynan and Wadi Ghuwayr (Barker et al. 1997, 1998; Hunt et al. 2004), and the fluvial systems of the southeastern Wadi Araba (Niemi and Smith 1999).

The increasing amount of research on alluvial fills and prehistoric occupations has resulted in two important aspects for landscape reconstruction: the association between terraces and locational patterns of prehistoric sites (Mabry 1992; Schuldenrein and Clark 1994; Cordova 1999a, 1999b; Macumber 2001) and the recurrence pattern of terraces and the climatic chronology of the Levant (Goldberg 1986; Macumber 2001). Despite all the recent research, the alluvial records of Jordanian wadis are still very scanty, especially because of sparse dating and because there are extensive areas have no records at all.

The majority of alluvial fills in Jordan consist of clastic deposits such as conglomerates, gravels, and sand, some of which are associated with ancient alluvial fans. Quaternary conglomeratic deposits are found along the Rift Valley (Bender 1974; Abed 1985; Macumber and Edwards 1997), in alluvial valleys dissecting the Western Highlands (Besançon and Hours 1985; Moumani, Alexander, and Bateman 2003), and around the large playas of the Central Plateau, such as Al-Jafr (Huckriede and Wiesemann 1968) and Azraq (Besançon and Sanlaville 1988; Copeland and Hours 1988; Besançon, Geyer, and Sanlaville 1989). Lower and Middle Paleolithic tools are often associated with conglomerates (Horowitz 1979; Baubron et al. 1985; Besançon and Hours 1985; Tchernov 1987; Macumber and Edwards 1997; Macumber 2001; Moumani, Alexander, and Bateman 2003). In part, this is because flint fragments found in these deposits offer excellent raw material for reduction cores.

Active alluvial fans are abundant in Jordan, especially in Wadi Araba, Wadi Rum, and Wadi al-Yutm in southern Jordan. As is evident elsewhere in the drylands of the world, alluvial fan deposits are good indicators of cycles of landscape stability and instability (Bull 1991). In Jordan, alluvial fan research has been implemented for the purposes of natural hazard assessment (Farhan, Beheiry, and Abu-Safat 1989), archaeological surveys (Niemi and Smith 1999), and research on neotectonic processes (Niemi et al. 2001). In general, in the early part of the Holocene, climatic fluctuations played a major role in fluvial landscape changes, whereas during the past five millennia, human agency has been a decisive factor (Goldberg and Bar-Yosef 1990).

Fluvial landscape change events recorded in wadis near archaeological sites are associated with important changes in cultural development. Some of these events are often found to be synchronous in various regions of the Levant. For example, a regional pattern of alluvial fill formation found in several geoarchaeological studies in the Levant is the formation of a terrace bearing Epipaleolithic materials (Goldberg and Bar-Yosef 1990; Macumber 2001). In Transjordan, this terrace seems to be recurrent in various wadis, particularly those dissecting the Western Highlands (fig. 5.5).

The representation of the terrace associated with Epipaleolithic materials and more recent fills is visible in the wadis of the Madaba-Dhiban Plateau (fig. 5.6). The Thamad Terrace, formed by a series of reddish brown alluvial and colluvial sediments, contains a sequence of Middle and Late Epipaleolithic occupations (Cordova et al. 2005). These sedi-

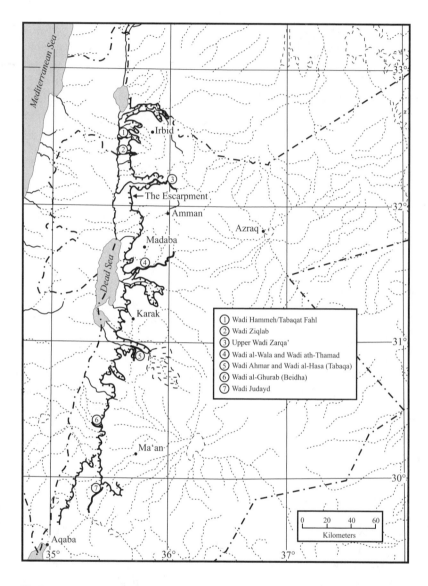

Figure 5.5 Terminal Pleistocene alluvial terraces associated with Epipaleolithic occupations in western Jordan.

ments were incised by stream erosion sometime in the Early Holocene, perhaps during the Pre-Pottery Neolithic; the first settlement following the incision dates to the Pottery Neolithic (Cordova et al. 2005). Subsequent aggradation occurred in the Middle Holocene through the formation of the Tur al-Abyad/Iskanderite Alluvial Unit, which was incised to form the Tur al-Abyad Terrace (fig. 5.6). This terrace is associated with Chalcolithic and Early Bronze occupations. The youngest and lowest historical terrace corresponds to alluvial deposits bearing Roman to Early Islamic materials.

A phase of alluvial accumulation during the Middle Holocene seems to have been recurrent across the wadis of the southern Levant (fig. 5.7). Associated artifacts and a few radiocarbon dates have placed this alluvial phase sometime between 6.5 ka and 5 ka BP (table 5.2). The high concentration of Early Bronze settlements along streams suggests that the floodplain formed by this alluviation phase was farmed and probably irrigated (Mabry 1992; Rosen 1995). The floodplain, however, did not last long. A widespread phase of stream incision eroded most of the sediments in the studied wadis sometime before four thousand years ago (table 5.2).

The best example of the sequence of events at a local scale can be shown in the case of Wadi al-Wala at the Early Bronze settlement of Khirbet Iskander (fig. 5.8). This erosional event seems to have had tremendous consequences for the farming communities of the Early and Middle Bronze Age (see chapter 6). This example shows the importance of documented alluvial stratigraphy sequences in the reconstruction of paleoclimatic events, paleohydrological changes, and human-environment relationships over time.

Lacustrine Deposits and Terraces

The lacustrine deposits of the Dead Sea and its precursors, Lake Lisan and Lake Samra, have been studied mostly on the western side of the Rift Valley (see Quaternary geology section in chapter 2). However, the numerous exposures visible along the eastern margin of the Rift Valley present multiple opportunities for correlation with the westside sections. Most of the lacustrine stratigraphy studies pursued in the Rift Valley have principally focused on the Lisan Marls (e.g., Bender 1974; Abed and Helmdach 1981; Niemi 1997; Abed and Yaghan 2000; Landmann et al. 2002; Edwards et al. 2004). The variability of sedimentary facies across

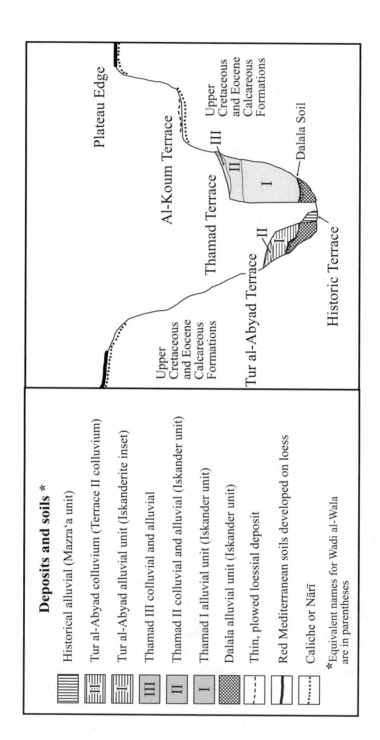

Figure 5.6 Geomorphic position and chronology of terraces and alluvial fills in Wadi ath-Thamad and Wadi al-Wala (Cordova et al. 2005:fig. 5).

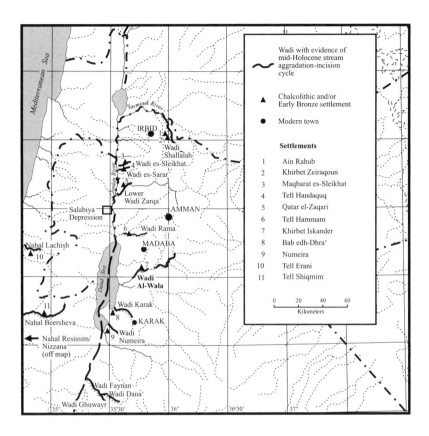

Figure 5.7 Reported alluvial fills associated with Chalcolithic–Early Bronze Age occupations. See table 5.2 for sources.

the basin and the poor dating of various members of the Lisan Formation represent an enormous challenge for interpreting the evolution of former Lake Lisan. The deposits of Lake Samra (Lake Amra) have been also identified in Jordan (Abed and Helmdach 1981), but the extension, levels, and ages of this lake are not known.

On the western margin of the Rift Valley, in the Bet Shan area, a close association between tufa deposits and paleolake levels was found (Kronfeld et al. 1988). Therefore, the potential for lake-level reconstruction in this area also exists on the Jordanian side. In the Tabaqat-Fahl area, in northwestern Jordan, researchers obtained relevant information from tufas for explaining the recession of lake levels and the transition from

Table 5.2 Mid-Holocene alluvial fill reported in selected wadis of the southern Levant

Stream[a]	Nearby Chalco-EB sites	Alluvial aggradation, 7–5 ka BP	Stream incision, ca. 4 ka BP[b]	References
Wadi Shallalah	Khirbet Zeiraqoun, Ain Rahub	Unit I	X	Cordova 2005b
Wadi es-Sleikhat	Maqbarat es-Sleikhat	H6 Unit	X	Mabry 1992
Wadi es-Sarar	Tell Handaquq	H6 Unit	X	Mabry 1992
Lower Wadi Zarqa[a]	Qatar el-Zaqari	H6 Unit	I	Mabry 1992
Wadi Rama	Tell Hammam	H6 Unit	I	Mabry 1992
Wadi al-Wala	Khirbet Iskander	Iskanderite Alluvial Inset	X	Cordova et al. 2005
Wadi al-Karak	Bab edh-Dhra[c]	Unnamed	X	Donahue 1981, 1985; Donahue et al. 1997
Wadi Numeira	Numeira	Unnamed	X	Donahue 1981, 1984
Wadis Faynan/Dana/Ghuwayr	Various Chalco-EB settlements	Faynan Unit	I	Barker et al. 1997, 1998; Hunt et al. 2004
Salibiya depression		Fz III	I	Schuldenrein and Goldberg 1981
Nahal Lachish	Tell Erani	Unnamed	X	Rosen 1986, 1995, 1997a
Nahal Beersheva	Shiqmim	Chalcolithic fill	X	Goldberg 1986
Nahal Resissim/Nizzana		Unnamed	X	Cohen and Dever 1981; Goldberg and Bar-Yosef 1990

[a] See figure 5.7 for locations.

[b] X = reported event; I = inferred or possible event.

Note: The abbreviation "ka BP" denotes the number of thousands of years before the present. "Chalco-EB" refers to Chalcolithic–Early Bronze settlements.

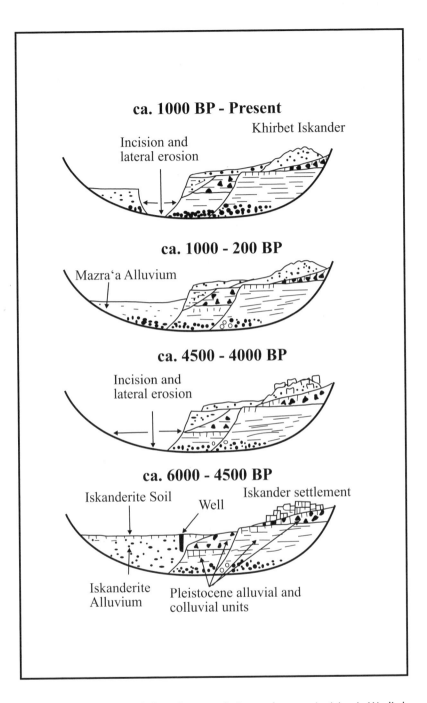

Figure 5.8 Sequence of alluvial accumulation and stream incision in Wadi al-Wala near the Early Bronze site of Khirbet Iskander.

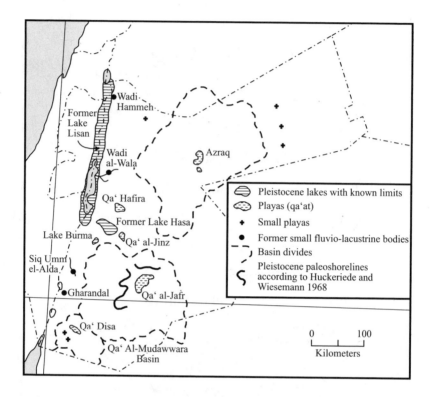

Figure 5.9 Location of Quaternary lacustrine deposits, landforms, and localities mentioned in text.

lacustrine to fluvial environments (Macumber and Head 1991). Farther south, in the Wadi Zarqa' Ma'in area, the potential of tufas for paleo-hydrological reconstruction has been explored (Khoury, Salameh, and Udluft 1984; Banat and Obeidat 1996).

Evidence of Pleistocene high level stands on elevated benches and terraces has been reported from around the playa (qa') basins of the Central Plateau (fig. 5.9). Qa' al-Jafr is the largest of the playa basins in Jordan; high paleoshore surfaces with Levalloiso-Mousterian lithic material indicate a high lake stand and consequently the presence of an extensive lake there during the Late Quaternary (Huckriede and Wiesemann 1968). In the Qa' al-Mudawwara basin, south of Al-Jafr, U/Th dating on materials from high paleobeaches at two locations yielded dates of 116 ± 5.3 ka BP and 76 ± 8.2 ka BP (Abed et al. 2000), which indicate high lake levels during the interglacial MIS 5e and during the glacial interstadial corresponding to MIS 5a (see fig. 5.1). Episodes of high lake level in the Jordanian desert have also been correlated with those of other basins in the Saharan North Africa and the Arabian Peninsula (Abed et al. 2000; Davies 2000).

Using geochemical properties from sediment cores from Qaʿ al-Jafr, Caroline Davies (2000) attempted to reconstruct paleolake history on the Central Plateau. Sediments in playa lakes (unlike those of lakes permanently occupied by water) are often exposed to wet and dry cycles, which affect the integrity of the sedimentary record. When the lakes dry out, deflation removes the exposed sediments, consequently creating stratigraphic gaps. These dry and wet cycles are the main reason that pollen grains are not preserved there (Davies 2000). When pollen grains are exposed to dryness, they are destroyed by weather elements or by shrinking clay.

Numerous smaller qaʿ basins are found in some localities in northern Jordan (Awawdeh 1998), the narrow valleys of Wadi Rum (Henry et al. 2003), and the lowest parts of the southern Wadi Araba (Abed 1998) (fig. 5.9). Despite stratigraphic gaps created by deflation, the deposits of some of these small basins may provide interesting information on paleoenvironments for certain periods during which accumulation was greater than today.

Pleistocene lacustrine deposits are also found in several areas of the Western Highlands. The largest of such lacustrine systems corresponds to Lake Hasa, which is known in Jordanian prehistory for its Pleistocene lakeshore occupation sites (Olszewski and Coinman 2002). Evidence for the existence of this lake is found in the relatively broad distribution of lacustrine marls in the area (Schuldenrein and Clark 1994; Moumani, Alexander, and Bateman 2003). Other Pleistocene lacustrine basins such as Qaʿ al-Jinz and Lake Burma are also linked to the former Lake Hasa basin (Davies 2000; Moumani, Alexander, and Bateman 2003).

Several smaller lacustrine bodies are found in valleys dissecting the plateaus of the Western Highlands (fig. 5.9). Some of them contain associated artifacts, as is the case of the Al-Wala Silts Unit on the Madaba-Dhiban Plateau (Cordova et al. 2005), and Wadi Siq Umm el-Alda (Schyle and Gebel 1997) and Wadi Gharandal (Henry et al. 2001) in southern Jordan.

The sparse radiometric dating of lacustrine deposits in Jordan results in a weak correlation between lake levels and climatic events of regional and global scale. Additionally, our knowledge of former shorelines in the playa lakes of the Central Plateau is scanty, despite the immense amount of evidence of prehistoric occupation on topographic benches around these former bodies of water.

Eolian Deposits

Despite their abundance in Jordan, eolian deposits have received little attention from geoscientists and archaeologists. In particular, loess is perhaps the least studied of all the Quaternary deposits in Jordan, despite its potential for landscape reconstruction as shown through studies in other areas of the Levant.

Donald Henry (1997a) and his colleagues (Henry et al. 2003) prompted an approach to eolian environments and prehistoric occupations in Wadi al-Hisma and Wadi Rum, where prehistoric occupations are separated by discrete mantles of silt and sand. Accordingly, sequential accumulations of silt and sand denote wet and dry periods, respectively (Henry 1997a). The logic of such a dichotomy is often associated with the relative amount of atmospheric moisture. For example, silt accumulation (or loess formation) in the northern Negev correlates with periods in the Pleistocene when spring storms were frequent and silt was available because of the amount of material removed from the mountains in the Sinai (Bruins and Yaalon 1979). In terms of atmosphere dynamics, this phenomenon can be explained as a result of a southward shift of cyclonic storms from the Mediterranean during the cold periods of the Pleistocene (Issar 2003). Accordingly, these cyclonic systems intensified dust-storm activity as their trajectories crossed the Libyan, Sinai, and Negev deserts, hence prompting silt accumulation (or loess accumulation) in the southern Levant. In contrast, dry periods and lack of storms meant a lack of fine silt deposition, under which circumstances, strong and dry winds mobilized massive amounts of sands (Bruins and Yaalon 1979; Goring-Morris and Goldberg 1990). This model suggests that the phases of loess deposition in the Negev occurred during the wet cycles and the incursion of sand dunes from the west during the dry cycles. To a large extent, Henry's (1997a) sequence of yellow silt and red sand correlates with wet and dry events of the Upper Pleistocene and Holocene (far right column in fig. 5.1).

Stable Isotopes

Carbon and oxygen stable isotopes have provided valuable information on past rainfall patterns. They are obtained from a variety of sources, including cave speleothems, microorganisms in lake- and sea-bottom deposits, land snails, and teeth (Goodfriend 1999).

Carbon stable isotopes are indirect indicators of rainfall through the amount of ^{13}C absorbed by plants. Low $\delta^{13}C$ values refer to C3 plants, which include mostly wooded species and cool-season grasses, while high values reflect an abundance of C4 plants, which include warm-season grasses. Thus, high amounts of low $\delta^{13}C$ values reflect larger amounts of rainfall because C3 plants, which absorb more ^{13}C during their photosynthetic process, are more abundant. Conversely, high $\delta^{13}C$ values reflect lower amounts of precipitation because C4 plants, which absorb less ^{13}C, are more abundant than C3 plants.

Carbon isotope studies are commonly performed on organic horizons in paleosols, but speleothems in caves usually produce interesting results. Water percolating down through cracks and pores has been enriched with carbon from the soils above the cave. When water precipitates carbon and other minerals in the cave, the precipitated minerals usually occur in layers along the surfaces of stalagmites and stalactites. These layers are the speleothems, which can be dated and arranged chronologically.

Oxygen isotopes are also frequently used for obtaining paleoclimatic proxies. They are likewise expressed in the form of a ratio, in this case between isotopes ^{18}O and ^{16}O, namely, $\delta^{18}O$. Their variations are related to the isotopic signature left by rainfall. For instance, rain that originated in the Mediterranean Sea has a different composition than rain that originated in the Indian Ocean. Therefore, $\delta^{18}O$ values have been used to determine relatively humid periods created by an increase of summer rain resulting from the enhancement of the Indian Ocean Monsoon. Although the analogies above seem simple, the mechanism of interpreting stable isotope data is more complicated because of a series of local factors and sources of carbon and oxygen.

In the Lisan Peninsula, a sequence of $\delta^{18}O$ values from the Lisan sediments encompassing the millennia between the Last Glacial Maximum and the Younger Dryas was obtained (Edwards et al. 2004). This sequence shows the alternation of humid and dry spells that were significant for the development of the Epipaleolithic cultures that culminated in sedentarization.

Both carbon and oxygen isotopes have been obtained from carbonate nodules formed in alluvial and eolian deposits in southern Israel (Magaritz, Kaufman, and Yaalon 1981; Goodfriend and Magaritz 1988), but because carbon is highly variable, it could not be used for paleoclimate reconstruction. However, oxygen isotopes are more uniform, especially in hard carbonate nodules developed in a carbonate-free sediment matrix (Goldberg 1994).

Stable isotopes from cave speleothems also provide paleoclimatic information of both local and regional significance. In Soreq Cave, located in the Judean Highlands near Jerusalem, researchers obtained $\delta^{18}O$ values (Bar-Matthews et al. 1998, 1999). In the same cave, numerical dates from the speleothem growth layers were then calculated using $^{230}Th/^{234}U$ series (Kaufman et al. 1998). These studies contributed to the establishment of a chronology of climate change during the Upper Pleistocene (Bar-Matthews, Ayalon, and Kaufman 2000) and Holocene (Bar-Matthews, Ayalon, and Kaufman 1998, 2000; Bar-Matthews et al. 1999). Oxygen isotope values from the speleothems in caves correlate with those in sapropel layers from sediments in the eastern Mediterranean Sea. Low-$\delta^{18}O$-value stages I, III, IV, and V in speleothems of Soreq Cave correlate with the formation of sapropels 1, 3, 4, and 5, respectively (fig. 5.1). Sapropels 1 and 5 correspond to humid periods with both winter and summer rains, while 3 and 4 represent high winter precipitation (Bar-Matthews, Ayalon, and Kaufman 2000). Of these, S5 (124–119 ka BP) and S1 (8.5–7 ka BP) show isotopic signatures indicating frost-free winters and drought-free summers, suggesting that enhancement in both the Mediterranean and the Indian Ocean Monsoon brought rain to the region (Bar-Matthews, Ayalon, and Kaufman 2000). Data for the northwestward expansion of monsoonal rains are also recorded in $\delta^{18}O$ records in cores in the Red Sea (Hembleben et al. 1996). However, controversy over the northward expansion of the monsoon into the southern Levant during the Early Holocene arises from surface paleosalinity obtained from the Dead Sea (Arz et al. 2003). Therefore, this issue is not yet fully supported, so more evidence from continental sediment records is needed.

Other caves with stable isotope records include the Ma'ale Efrayim Cave, located in the rain shadow of the Judean hills (Vaks et al. 2003), and numerous other caves in northern Israel (H. Geyh, reported in Issar

1990). Isotope patterns found in cave speleothems and marine sediments (Issar 2003) have been correlated with both oxygen and carbon isotopes obtained from the sediments of the Sea of Galilee (Lake Kinneret) (Stiller et al. 1983–84).

Potential sources of carbon and oxygen isotopes exist in various regions of Jordan. They include carbonate nodules developed in noncarbonated sediments, speleothems in caves in the limestone plateaus and highlands in western Jordan, and lacustrine sediments. The advantage of these studies, if implemented in Jordan, is that they can be correlated with the sequences west of the Rift Valley, including those from the bottom sediments of the eastern Mediterranean Sea and Red Sea.

Age Determination and Regional Cross-Correlation Issues

Our knowledge of landscape change sequences is based in large part on age-determination techniques. Relative dates based on sequences of archaeological materials are basic, especially where artifacts exist. Nonetheless, the use of absolute (numerical) dating techniques is fundamental in obtaining time resolution. The high costs of absolute techniques and the difficulties presented by the particular aspects of the Jordanian landscape have hindered the construction of high-resolution chronologies of sedimentary deposits for landscape reconstruction. As a consequence, the lack of age determinations has also reduced the possibility of cross-correlation of landscape alteration events.

The lack of organic matter in most Quaternary sedimentary deposits in the arid lands of the Near East reduces the capabilities of radiocarbon dating (Head 1999). Thermoluminescence (TL) and optically stimulated luminescence (OSL) are methods with great potential for dating eolian deposits. OSL also allows the dating of a series of other deposits, including alluvium. For this reason, TL and OSL are two of the alternative methods for dating Upper Pleistocene and Holocene deposits in arid lands. R. Neil Munro, Richard Morgan, and William Jobling (1997) applied OSL to date eolian sands in the area of Qaʿ Disa in southern Jordan (for location, see fig. 5.9; for aerial view, see fig. 2.4, bottom). They obtained six dates for a period spanning the past 200,000 years. Although the date ranges were too long for the dating of paleosols and periods of eolian deposition, the study showed the potential of this technique in eolian sediments. Additionally, Munro, Morgan, and Jobling (1997) discussed the potential of OSL dating for other deposits in Jordan. These include not only desert deposits but also paleosol horizons on the Western Highland plateaus (i.e., the red Mediterranean soils), land-slip deposits with sandy fills, such as the ones in Al-Baqaʿa Valley, sterile sandy infills in archaeological sites, and sandy fills in human-constructed agricultural terraces and dams.

The time frame to which each technique can be applied is another limitation that can hinder the possibilities of studying sequences of events spanning long periods of time. Radiocarbon dating is good only for ages younger than 50,000 years BP. Potassium-argon (K-Ar) dating can be done only on deposits and rocks of volcanic origin and cannot be done for recent ages. This limitation, however, is easy to manage because of the high variety of methods for different time ranges.

Thermoluminescence and electron spin resonance (ESR) dating have also been used for periods beyond the reach of radiocarbon dating. Although radiocarbon dates can be obtained for materials as old as 50,000 years BP, dates older than circa 30,000 years BP are often not accurate (Van Andel 1998a; Van der Plicht 1999). This has been an issue especially for prehistorians studying the 30,000–50,000 BP time frame, which encompasses the transition from the Middle Paleolithic to the early Upper Paleolithic (ca. 50,000–45,000 years BP), a critical time span for understanding the coexistence of modern humans and Neandertals and the demise of the latter. In view of the doubts about the use of radiocarbon dating for this period, TL has helped determine whether post-30,000 BP radiocarbon dates obtained from cave deposits are correct (Bar-Yosef 2000).

Another common problem in geochronology is the difference in results provided by different techniques. ESR measurements have been mainly obtained from animal teeth in various prehistoric contexts in the Levant. However, like radiocarbon dating, ESR may be limited by preservation issues; the poor preservation of dental remains typical of deposits in Jordan may hinder the possibility of widely using this technique for dating off-site Quaternary deposits. TL dates are obtained from quartz grains in sediments associated with findings. Although some come from the same stratigraphic layer, sometimes they are off, perhaps by several tens of thousands of years (fig. 5.1). Despite these problems, the application of ESR and TL is a major advance in establishing a chronology of hominid and early human habitation in the Near East.

Regional cross-correlation of ages is often a relevant topic in Quaternary science, because it permits determining whether paleoenvironmental events have local, regional, or global dimensions. However, the main limitation for regional cross-correlation of events is the scarcity of numerical dates, which is characteristic of vast areas in the Near East, including Jordan. Despite this problem, Quaternary scientists and archaeologists have undertaken several efforts to correlate events across the Levant and the rest of the Near East.

During the past fifteen years, several studies have addressed problems of dating techniques and cross-correlation in the areas of Quaternary science and archaeology of the Near East. Ofer Bar-Yosef and Renee Kra (1994) compiled and discussed a series of studies dealing with various dating techniques, correlation of ages obtained through different methods, and interpretation of proxy data in the eastern Mediterranean region. More recently, Hendrix Bruins, Israel Carmi, and Elisabetta Boaretto (2001) published a collection of papers dealing with various topics concerning radiocarbon-dating control in Near Eastern archaeology, especially on the issue of synchronization of historical and numerical dates and the correlation between archaeological and paleoenvironmental dates.

Bruins (2001) points to the dichotomy existing between radiocarbon dating for archaeological/historical chronologies and paleoenvironmental research. This dichotomy inevitably exists, as archaeologists and paleoenvironmental scientists obtain dates from different sources and for different purposes. On the one hand, archaeologists, who often obtain their dates from cultural deposits, have the main goal of dating lithic industries and assemblages, ceramic styles, and architectural features. On the other, geologists, geomorphologists, and similar specialists obtain their ages largely from natural deposits with the purpose of dating climatic, ecological, and geomorphological events. Because archaeological deposits have more datable material for methods such as radiocarbon dating and electron spin resonance, cultural chronologies have higher time resolution than do climatic and environmental sequences.

The unbalanced relation between archaeological and paleoenvironmental chronologies becomes evident when scholars try to factor in the influence of climate change on the development of cultural events. This situation is exemplified in the search for possible causes for the rise and collapse of the Early Bronze Age in the Levant (see chapter 6). In this case, a somewhat tight chronology of ceramics and architectural styles exists to explain growth and decline of urban life during the Early Bronze

Age (Philip 2001), but the poor resolution of most paleoclimatic change fails to provide a detailed sequence of atmospheric events that occurred at the centurial scale.

A similar situation arises when studying the Neolithic period. Tight chronological dating of cultural deposits places lithic technology and faunal and floral remains in a sequence that explains processes of animal and plant domestication, but this sequence bears no parallel with the low-resolution paleoenvironmental record. For instance, the decline of settlement at the end of the Pre-Pottery Neolithic B, which has often been attributed to climatic and human-induced environmental deterioration, has little resolution in pollen and isotope records or other sedimentary records.

Although the dichotomy between archaeological and paleoenvironmental dating is initially inevitable, this situation can be ameliorated if all the specialists involved discuss their results and find better ways to combine dating work. This requires the organization of multidisciplinary teams with specific objectives and the scheduling of frequent workshops and meetings to present and discuss data. This issue perhaps should encourage geoscientists to publish their results in archaeological journals and archaeologists in geoarchaeological and Quaternary science journals.

Calibration and correction of absolute ages are two significant issues that are often taken for granted. In some cases, the problem originates in a faulty understanding of what calibration and correction means. Most laboratories report both uncalibrated dates and calibrated dates; it is up to the researcher to use one or the other. Consequently, the reports and publications contain sometimes calibrated and sometimes uncalibrated dates. The problem arises when uncalibrated dates in one study are correlated with calibrated dates obtained in a different study. The best expression of this problem is exemplified in the aforementioned archaeological-Quaternary dichotomy of dating and correlation. While archaeologists prefer to report numerical ages in calibrated calendar years (AD or BC), most Quaternary scientists use uncalibrated dates BP or conventional radiocarbon years before the present. In some cases, however, when Quaternary scientists present calibrated dates, they do so as "cal. BP." In addition, there are numerous different ways of presenting dates, which vary from author to author and from journal to journal, despite calls for universal conventional ways of reporting dates (Colman et al. 1987).

The problem of calibration comes together with a series of misconceptions regarding the technical nature of ^{14}C dating, such as isotopic fractionation correction, calibration, the question of absolute dates, geochemical complications such as reservoir ages, single-year versus multi-year samples, and wiggle-match dating (Van der Plicht and Bruins 2001). These problems are rarely if ever discussed in papers in which dates are presented. Perhaps one way to solve this problem would be to publish tables providing both calibrated and uncalibrated dates, regardless of what the authors use in their interpretations. Thus, if calibrated calendar ages are reported, then they should be referenced to a table containing the original uncalibrated dates along with a summary of all the information provided by the lab. If uncalibrated dates are reported, then they likewise should be referenced with the calibrated calendar ages.

6
Patterns of Millennial Landscape Change

The sequence of climatic events and landscape change that occurred in Transjordan during the past 20,000 years has the potential to provide a sound foundation for cross-correlation of cultural developments and adaptation strategies. Unfortunately, local and regional paleoclimatic chronologies are still scarce. As additional geomorphological, paleontological, and archaeological studies are conducted, more information can be added to fill out this scanty body of knowledge. However, as the discussions in earlier chapters indicate, paleoclimatic reconstruction and the dating and analysis of cultural developments in Jordan pose numerous challenges for researchers throughout the Paleolithic and into more recent eras.

The Pleistocene in Jordan: A Brief Introduction

Traditionally, the Pleistocene period has been subdivided into the Lower, Middle, and Upper Pleistocene (see fig. P.2, preface; Williams et al. 1998). However, these divisions are seldom used in the literature dealing with climate change, in which a division into marine isotopic stages (MIS) is preferred (see fig. 5.1). Unlike the high latitudes of Eurasia and northern North America, where glacial, fluvioglacial, and loess deposits have been studied comprehensively, the Levant has yielded little information useful for generating a paleoclimatic chronology of the Pleistocene. Attempts have been made to match local climatic changes with European chronologies (e.g., Horowitz 1979; Leroi-Ghouran and Darmon 1991), but strong physico-geographical differences between the two regions make any kind of paleoclimatic correlation difficult. Unlike Europe, the Levant occupies a drier area that is impacted by subtropical and tropical atmospheric circulation systems. In addition, the Levant is affected by different amounts of solar radiation. Rainfall in the Levant is heavily influenced by the Mediterranean Sea and occasionally by northward movements of the Intertropical Convergence Zone (ITCZ), creating a flow of moisture through the enhancement of the Indian Ocean Monsoon

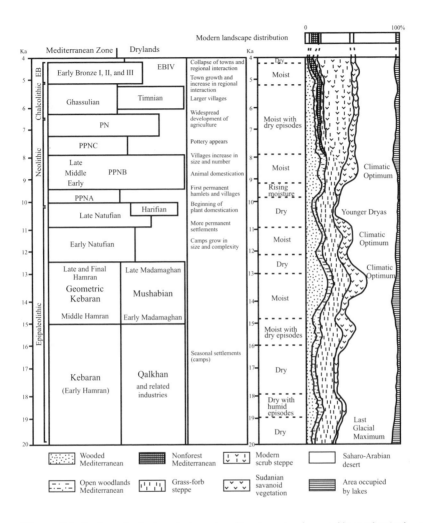

Figure 6.1 Archaeological periods, paleoclimate episodes, and hypothetical fluctuations of the area occupied by vegetation regions for the period 20,000–4,000 years ago. Cultural chronology is after Henry 1997b, with modifications and additional data.

(Roberts and Wright 1993; Blanchet, Sanlaville, and Traboulsi 1997). Consequently, Quaternary scientists have opted for establishing a regional paleoclimatic chronology. Thus, there has been a trend toward referring to climatic chronologies based on isotopes and sapropels in the eastern Mediterranean Sea and lake-level fluctuations (see fig. 5.1).

From the point of view of prehistoric archaeology, the Pleistocene is divided into the Lower (1.8 Ma–125 ka BP), Middle (125–50 ka BP), and Upper Paleolithic (50–20 ka BP) and the Epipaleolithic (20–10 ka BP). These broad periods encompass a series of stone tool technologies developed by hominids and modern humans. Lower Paleolithic stone tools in the Ubeidiya Formation in Israel are the first evidence of *Homo erectus* in the Levant (Tchernov 1988). Evidence of Lower Paleolithic hominid presence in Jordan exists in many locations. For example, lithics and animal bones have been reported in deposits in high terraces of Wadi Zarqa' contemporaneous with those of the Ubeidiya Formation (Kafafi et al. 1997; Parenti et al. 1997). Acheulian (Lower Paleolithic) materials associated with conglomerates and tufa deposits have been reported in the Tabaqat Fahl Formation in the Wadi Hammeh area (Macumber and Edwards 1997). Likewise, conglomerate deposits elsewhere along the Rift Valley (Macumber and Edwards 1997) and in the Azraq region (Rollefson et al. 1997), Al-Jafr basin (Quintero, Wilke, and Rollefson 2002), and southern Jordan (Henry 1995) have yielded evidence of Lower Paleolithic hominid activity. However, because of intense and long-spanning geomorphological processes, most artifacts found in these contexts are unstratified (Copeland 1997).

Middle Paleolithic human remains have been recovered from caves along the Mediterranean Levant but not in Jordan. Whether the lack of hominid bones during this period is attributable to the drier and harsher environments of Transjordan or to the paucity of surveyed and excavated localities remains unclear. However, the recent rapid increase in Paleolithic research in Jordan will probably succeed in producing direct evidence of hominids and occupation sites in this part of the Levant. Thus far, rockshelters in Tor Faraj and Tor Sabiha in the south (Henry 1995, 2003b) and Ain Difla in Wadi al-Hasa (Clark et al. 1997) have provided interesting cultural and paleoenvironmental information for the Middle Paleolithic and occupation presumably by Neandertals.

The end of the Riss glaciation in the Alps and the Illinoian glaciation in North America corresponds to MIS 6. The subsequent interglacial period corresponds to MIS 5e (125–105 ka BP) (see fig. 5.1). Globally, during the MIS 5e, temperatures were somewhat higher than in the Holocene, which is also considered an interglacial period. Unfortunately, no data on paleotemperatures and paleoprecipitation exist for the Levant. Some data from the adjacent seas, however, support the global data, suggesting that regional conditions were warmer and more humid.

Pollen, phytoliths, and faunal remains from the Middle Paleolithic deposits of Tor Sabiha and Tor Faraj in southern Jordan suggest more humid conditions than today's climate exhibits (Emery-Barbier 1995; Henry 1995, 1998a, 2003b; Rosen 2003). The existence of high beaches bearing Levalloiso-Mousterian lithics in Al-Jafr basin suggests higher lake levels during parts of the Middle Paleolithic (Huckriede and Wiesemann 1968). U/Th dates obtained in the Qaʿ al-Mudawwara basin (Abed et al. 2000) indicate that a high lake level occurred in the region at 116 ± 5.3 ka BP (MIS 5e). Likewise, this period witnessed the filling up of Lake Burma, in the upper reaches of the Wadi al-Hasa drainage basin (see fig. 5.1) (Moumani, Alexander, and Bateman 2003). The high lake levels during these stages are attributed to an increase of moisture produced by enhancement of both winter Mediterranean and summer Indian Ocean monsoonal rains (Abed et al. 2000; Bar-Matthews, Ayalon, and Kaufman 2000; Davies 2000). MIS 5d, 5c, and 5b were cooler and drier than MIS 5e. MIS 5a was again a humid period, corresponding to a warm stage dividing the two main cold phases of the Würm glacial period. Again, pollen evidence suggests ameliorated conditions (Emery-Barbier 1995), and lake-level stands seem to be higher (see fig. 5.1). Another high level stand has been recorded in the Qaʿ al-Mudawwara basin at 76 ± 8.2 ka BP (Abed et al. 2000).

MIS 4 (70–60 ka BP) corresponds to a relatively cool period along the transition between early and late stages of the Würm glaciation. Oxygen isotopes in the oceans mark a rather cool period. There are no detailed records for this period in Jordan. However, lakes are assumed to have receded. The transition from Lake Samra to Lake Lisan occurred during this time. In the Lake Burma deposits, a calcrete dated 57 ± 4 ka BP and the conglomerate below it may indicate a dry-out event (Moumani, Alexander, and Bateman 2003). Farther south, the sequence of eolian deposition in Jebel Qalkha (Henry 1997a) indicates dryness. In Jordan this cool period may have been represented by dryness, but there is not enough evidence to support this assumption, particularly during the entire duration of MIS 4.

MIS 3 (60–25 ka BP) represents a series of alternating cooling and warming phases. Changes in the levels of Lake Lisan are relatively low but stable (see fig. 5.1). Thus, between 60 ka and 30 ka BP, Lake Lisan stabilized at a low level between −280 and −290 meters, with a single drop to −360 meters around 47–48 ka BP (Bartov et al. 2002). Afterwards, the level of Lake Lisan increased, with a maximum level recorded at −164 meters around 27 ka BP (Bartov et al. 2002). Lake Hasa also increased in size, becoming substantial between 25 ka and 19.5 ka BP (Schuldenrein and Clark 2001). Lake Burma still existed, although it probably began to recede by 20 ka BP (Moumani, Alexander, and Bateman 2003). The demise of the Neandertals in the Middle East and Europe and the dominance of anatomically modern humans (i.e., *Homo sapiens*) occurred during this stage. However, the influence of climate change in the development of these two events is still a matter of debate among prehistorians.

MIS 2 (25–15 ka BP) corresponds to the last advance of the world's continental ice sheets before the glaciation was over. For the Levant, during this last glacial period, cooling meant humid conditions at the beginning (24–20 ka BP), followed by an extremely cold and dry phase (20–16 ka BP) (Sanlaville 1996; Henry 1997b). However, data in certain areas suggest that this was not necessarily true for the entire Levant (Wigley and Farmer 1982; Van Zeist and Bottema 1991; Baruch and Bottema 1999). As illustrated below, the proxy data for this period exhibit contradictory information.

The Last Glacial Maximum and the Terminal Pleistocene

Paleoclimatic Reconstruction

The Last Glacial Maximum (peaking around 20 ka BP) coincides with the end of the Upper Paleolithic and the beginning of the Epipaleolithic period. The latter is subdivided into Kebaran, Geometric Kebaran, and Early and Late Natufian, which are in turn subdivided into specific lithic industries and assemblages (fig. 6.1). These subdivisions vary regionally, suggesting technological adaptations to the different environments of Jordan and the Levant (Henry 1997b; Olszewski 1997; Clark, Coinman, and Neeley 2001). The Epipaleolithic period is also divided into Early (20–14.5 ka BP), Middle (14.5–12.5 ka BP) and Late (12.5–10 ka BP) Epipaleolithic (Byrd 1998). This division aims to distinguish broad cultural changes that coincide with major climatic episodes.

During the coldest phase of the Last Glacial Maximum (20,000–16,000 years ago), vegetation in the mountainous regions of Turkey and Iran was characterized by *Artemisia* steppe, while the Mediterranean coasts of Turkey and Greece were characterized by open oak woodlands

(Van Zeist and Bottema 1991). Pollen sequences in the Levant present conflicting pictures during this period. Originally, the Ghab Valley records pointed to an increase of arboreal vegetation during full glacial times (Niklewski and Van Zeist 1970), which led to the belief that the coldest stage of the glaciation was not so dry in the Levant as in the rest of the Near East. However, this pollen sequence rendered poor time resolution because of sparse dating. One new pollen diagram from the Ghab Valley (Yasuda, Kitagawa, and Nakagawa 2000), although tightly dated, did not include the Last Glacial Maximum (LGM) in the sequence. The pollen zone at the bottom, around 15 ka BP, shows a transition from steppe vegetation to a forested environment, which is more compatible with pollen diagrams in Turkey and in the southern Levant.

Disagreement exists among other sets of paleoclimatic records as well. For example, the levels of Lake Lisan were apparently higher during the LGM, contradicting the arid signatures provided by pollen records (see fig. 5.1). The high stands of Lake Lisan between 21 ka and 10 ka BP have been interpreted as the result of low temperatures, and consequently low evaporation rates, rather than an increase in precipitation (Stein 2001). This scenario makes sense in the context of global low temperatures but rules out other factors such as hydrological and geological aspects linked to the water sources of the lake. An important amount of water feeding the Dead Sea originates in springs and not directly from rainfall (Abu-Jaber 1998; Capaccioni et al. 2003). However, this possibility is difficult to explore, since proxy data for spring paleoflow do not exist.

In the Jebel Qalkha area, the Early Hamran (Kebaran) sites are found on drift sand that appears to have accumulated during the LGM (Henry 1998a). However, pollen data from the rockshelter deposits in the same region point to a relative increase in arboreal pollen (Emery-Barbier 1995). This has been interpreted as the result of short-term moist events toward the end of the LGM (Henry 1997b). Around the same time, palustrine conditions in the Wadi al-Hasa and a high incidence of arboreal riparian and aquatic pollen indicate climatic amelioration (Schuldenrein and Clark 1994). Signs of pedogenesis, which is a proxy for moister climate, have been reported in soils between 18,000 and 16,000 years ago, a time that is generally considered dry in the rest of the Levant (Garrard et al. 1988). Carbon isotopes from the Negev indicate that the C4:C3 ratio in plants during this time period was not much different than that of today (Goodfriend 1999; Bar-Yosef 1998), which suggests that conditions there were not extremely arid, or at least not drier than today.

It is not clear, however, why the now-drier areas of the southern Levant did not record extremely dry conditions during the LGM. A southward shift of westerly winds in the eastern Mediterranean region has been proposed (Wigley and Farmer 1982); this would have meant drier conditions for the northern parts of the Near East, as evidenced by pollen diagrams in Anatolia (Van Zeist and Bottema 1991) and Al-Ghab (Yasuda, Kitagawa, and Nakagawa 2000), and somewhat moister conditions in the southern Levant. Although mean annual rainfall may not have been abundant, frequent cloudiness and low temperatures could have kept evapotranspiration rates low, allowing more available moisture for plants. Under such conditions, deserts could have sustained steppe vegetation.

The warming of the atmosphere following the LGM resulted in increasing atmospheric moisture and consequent expansion of arboreal vegetation (Baruch and Bottema 1991, 1999). High arboreal pollen frequencies in the diagrams of the Ghab Valley (Yasuda, Kitagawa, and Nakagawa 2000) and Lake Hula (Baruch and Bottema 1999) indicate periods of increased moisture between 15 ka and 12 ka BP. Although moisture prevailed through most of the period between 15 ka and 11 ka BP, brief dry events occurred. Based on numerous archaeological and geoarchaeological data throughout the Levant, an alternating sequence of two climatic amelioration episodes and two dry episodes for the period following the LGM have been proposed (Sanlaville 1996; Henry 1997b). Accordingly, the moist event during 15–13 ka BP was followed by a cold-dry episode 13–12 ka BP. Then wet conditions returned between 12 ka and 10.8 ka BP, and this was followed by a brief cold-dry episode during the Younger Dryas, 10.8–10 ka BP. This sequence is evident in pollen diagrams obtained from cultural deposits in southern Jordan (Emery-Barbier 1995). The moist conditions of the two wet episodes resulted in an expansion of wooded areas into steppe and subsequently the expansion of steppe into desert areas (fig. 6.2).

The climatic episodes described above correlate with the curve of Lake Lisan levels (see fig. 5.1). The postglacial highest stand (at −180 meters) occurred sometime between 15,000 and 14,000 years ago (Yechieli et al. 1993), coinciding with the climatic amelioration of 15–13 ka BP. Conversely, the drop of lake levels to their minimum (−700 meters) (Begin et al. 1985) coincided with the short but very dry event known as the Younger Dryas.

Soil profiles developed in alluvial sediments also show evidence of

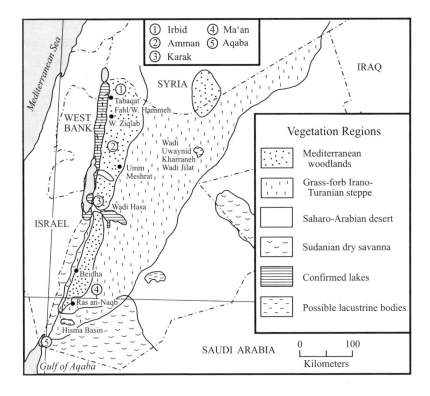

Figure 6.2 Epipaleolithic sites mentioned in text and possible expansion of woodlands and steppe around 12,500 years ago. Compare with the distribution of modern landscape regions in fig. P.1.

moist conditions between 16 ka and 12 ka BP. Paleosols in association with Middle Epipaleolithic occupations have been reported in the Wadi Hammeh/Tabaqat Fahl area (Macumber and Head 1991), Wadi Ziqlab (Maher and Banning 2002), Wadi Jilat (Garrard et al. 1988), and Umm Meshrat 1 (Cordova et al. 2005). Periods of alluvial accumulation in between paleosols correspond to short geomorphic destabilization events that occurred at the end of dry events when wet conditions returned swiftly. These cycles of stability and erosion result in the stratification of most Epipaleolithic occupations in alluvial settings. Examples of such settings exist in the deposits in Tabaqat-Fahl and Wadi Hammeh (Macumber 2001), Wadi Ziqlab (Banning et al. 1992; Field and Banning 1998), Wadi ath-Thamad and Wadi al-Wala (Cordova et al. 2005), Wadi al-Hasa and Wadi Ahmar (Byrd and Colledge 1991), Beidha in Wadi al-Ghurab (Field 1989), and Wadi Judayd (Hassan 1995). In some of these cases, sequences of Kebaran and Natufian occupations are found stratified in sequences of colluvial and alluvial deposits.

The wet-dry oscillations that occurred during the Terminal Pleistocene caused destabilization in the plateaus of the Western Highlands. During this time, large tracts of soils and sediments were stripped from the Northern Moab Plateau (Cordova 2000). The removal of red Mediterranean soils from the plateaus produced sediments that accumulated along the main valleys, which explains the reddish tones of the Thamad Alluvial Unit, dated roughly between 18 ka and 12 ka BP (Cordova 2000; Cordova et al. 2005). Evidence for intense accumulation of red clayey sediments also exists in the sediments of the Damya Formation on the top of the Lake Lisan Marls (Abed and Yaghan 2000). The reddish clay layers reported in the clastic materials of the Upper Member of the Lisan Formation around 13 ka BP (Yechieli et al. 1993) also suggest an erosional event of regional magnitude. These red clay deposits may correspond to the phase of intense erosion that affected the red Mediterranean soils on the plateaus on both sides of the Rift Valley.

Cultural Development

The complicated swing between moist and dry episodes that characterizes the Terminal Pleistocene resulted in a highly complex cultural sequence (Henry 1997b). An analysis of settlement patterns, industries, and other cultural aspects indicates that the Early Epipaleolithic populations were able to adapt to oscillations between dry and wet conditions, especially between 20,000 and 14,500 years ago (Byrd 1998). Conditions became more stable after 14.5 ka BP, for although the wet-dry cycles became longer, moisture levels remained relatively high in general.

The enhancement of lakes produced by increasing moisture augmented the availability of food opportunities, especially through hunting. For example, human populations around Lake Hasa benefited from the large numbers of Gazella sp. (gazelle), Equus sp. (horse), Bos sp. (wild cattle), Testudo sp. (tortoise), ovi-caprids, and a variety of birds (Clark et al. 1988). In Wadi Jilat, botanical and faunal assemblages suggest steppe conditions in an area that is now desert (Garrard, Colledge, and Martin 1996). Thus, it is evident through a series of data, especially on fauna, that wooded vegetation and grasslands extended farther east because of the optimal climatic conditions that prevailed throughout most of the Middle Epipaleolithic (fig. 6.1). These optimal environments all over the Levant are believed to have eventually produced the favorable conditions needed for sedentary life and farming (Baruch and Bottema 1991; Bar-Yosef and Valla 1991; Bar-Yosef and Meadow 1995; Hillman 1996; Henry 1997b).

Early and Middle Epipaleolithic occupation sites were associated with mobile hunter-gatherer economies, which attained some degree of specialization, as reflected in the variety of camps for each ecological zone (Byrd 1998). In southern Jordan, two groups have been identified (Henry 1995): the Hamran and the Qalkhan-Mushabian. The Hamran were mainly associated with the Mediterranean uplands of the Ras an-Naqb Plateau, while the Qalkhan-Mushabian were associated with the low and dry areas of Al-Hisma basin. The subsistence strategies of each of these two groups were adapted to their particular environments, although the populations coalesced in the lower areas during the winter. During the warmer season, the two groups dispersed into smaller bands over a larger territory that included parts of the Red Sea coast and the Rift Valley.

In Wadi al-Hasa, the pattern of settlement was focused on permanent sources of water (Clark et al. 1988). In the Wadi Jilat and Karraneh areas, settlements concentrated around periodic spring and oasis settings (Garrard, Baird, and Byrd 1994). However, there was a tendency to occupy large open-air base camps during the dry and cool times of the year (Byrd 1998). The variety of landscapes and the diversity of groups living in them resulted in a complex human mosaic on the Epipaleolithic landscapes, for which several interpretations have resulted in different models of human-environmental relations (see Schuldenrein and Clark 2001, 2003).

During the Late Epipaleolithic, the Natufian adaptive strategy represented a dramatic departure from the generalized mobile hunting and gathering of the previous periods (Henry 1989). Although archaeologists originally associated the Natufian with the Mediterranean climatic areas of the Levant, surveys in recent years have found numerous Natufian sites in the drylands, which were formerly considered marginal zones (Henry 1998b; Sellars 1998).

The main characteristic of the Natufian in the Levant is the shift to subsistence on wild cereals and more stable and wealthier settlements (Sellars 1998). The location of Natufian sites on the woodland-steppe ecotone and near permanent sources of water is perhaps one of the main environmental recurrent patterns of settlement during this period. This settlement pattern allowed the stable settlers to exploit woodland resources and open areas during longer periods of time during the year (Henry 1989). This situation seems to be consistent among Jordanian Natufian sites, as is evident at Beidha, Taibe–Ain Rahub, and numerous sites in Wadi Ziqlab, Wadi Hammeh, Wadi ath-Thamad–Wadi al-Wala, Wadi al-Hasa, and Ras an-Naqb. Those found near water sources in the steppe-desert ecotones include numerous sites in the Hisma basin, Wadi Jilat, Wadi Uwaynid, and the Black Desert (fig. 6.2). The use of the biotic richness commonly found in ecotones is one characteristic typical of the human-environmental relations and a subsistence strategy during the Natufian period.

Human Impacts on the Landscape

Although farming and pastoralism had not yet appeared, minimal impact on the landscape may have occurred through other human activities during the Epipaleolithic. First, the increase of semipermanent settlement in certain regions, especially during the Natufian period (12.5–10 ka BP), may have had an impact on flora and fauna around the sites. Anthropogenic plant communities, mainly weeds, may have formed in and around the settlements (McCorriston and Hole 1991; Wright 1993). The increasing population may have had an impact on vegetation through the increase of fires, although no study has proved the extent of such an impact.

The large number of ungulates, especially gazelle, found among faunal remains in Epipaleolithic sites implies intense hunting (Garrard et al. 1988; Byrd 1998). Macrofaunal remains of mammals indicating sedentism, such as rats (Tchernov 1991), as well as macrobotanical remains of weeds (Zohary 1973; Hillman 1996) indicate minimal, though evident, transformation of flora and fauna.

Direct human impact on soils was nonexistent because tilling for farming had not yet been implemented. Soil erosion occurred as a result of vegetation change, prompted by rapid climatic fluctuations. Intense soil erosion on the Madaba-Dhiban Plateau (Northern Moab) during the Epipaleolithic is evident in the massive accumulations of reddish brown silt that originated from the plateau soils. These accumulations form the Thamad Alluvial Unit of the Wadi ath-Thamad and its upper tributaries

(Cordova 2000). However, the causes of such dramatic erosional events are related to the rapid climatic changes resulting from the transition of global cold to warmer climates (Cordova 1999b, 2000).

The Younger Dryas as a Prelude to a Major Cultural Change

Paleoclimatic Reconstruction

A dry and cool episode of worldwide magnitude, the Younger Dryas was dated in the GISP2 ice core in Greenland as occurring between cal. 12.8 ka and 11.4 ka BP (uncalibrated 11–10 ka BP). The name "Dryas" comes from *Dryas octopetala*, a plant of tundra environments, whose pollen grain marks the return of cold conditions in northwestern Europe. But what was the climate like in the Near East during this event?

Pollen diagrams show in general a drop in arboreal pollen, suggesting dry and cold conditions. In eastern Turkey and Iran, where postglacial forests had not fully developed, this change was not as conspicuous as it was in western Turkey and Greece (Bottema 1995) or in northern Israel (Baruch and Bottema 1991), where forested areas had already increased between 15 ka and 12 ka BP.

In the Rift Valley, Lake Lisan dropped to its lowest-ever level about 11–10 ka BP (see fig. 5.1). At Tabaqat Fahl (Macumber and Head 1991), the depositions of fluvio-deltaic sediments associated with Epipaleolithic occupations mark the recession of Lake Lisan. Lake Hasa (Schuldenrein and Clark 2001) also dried out at the end of the Terminal Pleistocene, but the climatic causes of its drying are difficult to determine if one considers local tectonic factors. Joseph Schuldenrein and Geoffrey Clark (1994) point to faulting as one of the main reasons why the head of Wadi al-Hasa eroded backwards, eventually reaching the Lake Hasa basin. Nonetheless, backwards erosion could have reached the lacustrine basin when the lake was already dry. Geomorphological data from Wadi ath-Thamad and Wadi al-Wala show a clear shift to arid conditions, as silt alluvial deposition decreased and was replaced by coarse, angular gravels (Cordova et al. 2005).

Cultural Development

In the Levant, the Younger Dryas period coincides for the most part with the Late Natufian and its transition to the Pre-Pottery Neolithic period

(Bar-Yosef 1998). Therefore, the Younger Dryas coincides with a series of cultural changes that led to sedentary life and the beginning of farming. This coincidence, however, sounds ironic, as the Younger Dryas is known for its climatically adverse conditions.

Some scholars postulate that dry conditions stimulated plant cultivation among groups that were already in the process of sedentarization (Henry 1989; Bar-Yosef and Belfer-Cohen 1992; Moore and Hillman 1992). Other scholars disagree with this view and propose an alternative scenario in which dry conditions forced domesticable wild plants to migrate to areas where human populations were already manipulating wild plants to increase food production (McCorriston and Hole 1991). For this reason, the understanding of climatic change is crucial for understanding the Natufian period (Baruch and Bottema 1991; Bar-Yosef and Valla 1991).

The cultural ecological variability of the Natufian period can be appreciated in the diversity of landscapes that the studied sites of this period occupied. Thus, the original assertion that the Natufian culture was mainly an aspect confined to the Mediterranean region has been overturned by recent studies in the dry zones (Henry 1998b). Consequently, the trend has now turned to analyzing site function in the context of local environments rather than regional patterns. Unfortunately, the scarcity of paleoclimatic data reduces the focus on small-scale environmental developments. Nonetheless, interesting schemes of resource utilization have already been created for the areas around the former Lake Hasa (Schuldenrein and Clark 2001, 2003) and southern Jordan (Henry 1995).

The Early Holocene and the First Farming Societies

Paleoclimatic Reconstruction

Starting about 10 ka BP, temperatures rose and rains gradually increased to reach a maximum between 9 ka and 8 ka BP. The time span between 9 ka and 6 ka BP, a relatively humid period in the Levant, has been recognized as the Holocene Climatic Optimum (Blanchet, Sanlaville, and Traboulsi 1997; Rossignol-Strick 1998). The increase in precipitation is recorded in many parts of the eastern Mediterranean region (Rossignol-Strick 1999) and areas as far inland as the Zagros Mountains (Van Zeist

and Bottema 1991). This climatic optimum was the result of rising temperatures and the increase in cyclogenesis in the Mediterranean, as well
as the enhancement of the monsoon in the Indian Ocean. These changes
meant a substantial increase of winter rains and a possible increase of
summer rains in the Arabian Peninsula, Egypt, and the southern Levant (Fontugne et al. 1994; Blanchet, Sanlaville, and Traboulsi 1997; Bar-
Matthews et al. 1999). However, there are no paleoclimatic proxy records
that support the occurrence of summer rains during the climatic optimum anywhere in Jordan. Several records show evidence of moist conditions but not rain seasonality.

Pollen diagrams from lakes in the mountains of Iran and Turkey show
that woodlands and forests advanced into areas formerly occupied by
steppes (Van Zeist and Bottema 1991). But at the same time, the overall amelioration of the climate during this phase stimulated the prolif-
eration of farming villages and pastoralism. Thus, although forests and
woodlands expanded as a response to this moisture increase, their full
expansion was curbed by the increase in population and agricultural activities (Roberts 1982; Yasuda, Kitagawa, and Nakagawa 2000). A similar
scenario occurred in the drylands, where an increase in moisture favored
grassland expansion, consequently allowing more possibilities for farming and herding—activities that in the end had a detrimental effect on
the steppe and desert vegetation (Butzer 1995).

The big picture provided by regional paleoclimate records in the Levant shows overall warm and humid conditions, which are optimal for
the expansion of woody vegetation and for the development of farming.
Therefore, researchers often take for granted the obvious climatic optimum without paying attention to short-term climatic events that may
have had an impact on subsistence strategies of the Neolithic peoples.
In view of this issue, Arie Issar (2003) discusses changes in settlement
patterns, site size, and specialization in the context of climatic fluctuations detected by stable isotopic studies in the Levant. Even more serious
is the issue of continentality of the Transjordanian territory, where the
influence of the seas is diminished compared to the coastal areas of the
Levant, where most of the data have been generated.

A series of dated marine records points to climatic change on a
macroregional scale (Fontugne et al. 1994; Rossignol-Strick 1999). Stable
isotope data from lakes and caves in Israel constitute a more local
source of evidence, which correlates with sea-bottom proxy data (Bar-
Matthews, Ayalon, and Kaufman 2000; Issar 2003). Other proxy data

such as soil development, lacustrine deposits and terraces, and alluvial and eolian chronology show very low resolution for this period in Jordan. The scarcity of local paleoclimatic records for this period in Jordan has resulted in a trend toward using macroregional proxy data from neighboring regions. The evidence for a rather humid period during most of the period spanning 9–8 ka BP comes from faunal and floral macroremains from archaeological sites. However, these sets of data attest to rather local conditions around sites and thus give little information about rainfall abundance, seasonal rain patterns, and short-term dry events.

Cultural Development

The cultural period corresponding to the Early Holocene is the Neolithic, which has been subdivided into chronological phases based primarily on the presence or absence of ceramics and further subdivided on the basis of lithic technology and assemblages, architecture, subsistence economy, and ritual (table 6.1). The beginning of the Pre-Pottery Neolithic A (PPNA) phase occurred during the end of the cold and dry event of the Younger Dryas, but researchers have not yet determined how the transition from dry to moderately humid conditions influenced settlement patterns. Settlement may have expanded into dryland areas as the expanses of woodlands and steppes increased at the expense of the deserts (fig. 6.3). In view of the poor resolution provided by the extant climatic records for this period, the gradual improvement in moisture may be assumed to have influenced a slow increase in farming activities (Issar 2003). In Jordan, only five sites have been assigned to the PPNA: Sabra 1, Iraq ed-Dubb, Dhraʿ, Zahrat adh-Dhraʿ 2, and Wadi Faynan 16 (Rollefson 1998). A sixth one, Jebel Quaisa (J-24), is considered only partially PPNA (Henry 1997b). A seventh potential site is Zahrat adh-Dhraʿ 2, located in the Lisan Peninsula area, where excavations have yielded remains of domesticated barley, wheat, and pulses (Edwards et al. 2002).

The Pre-Pottery Neolithic B (PPNB) has yielded more paleoecological information because of the abundance of excavated sites in all regions of Jordan. Paleoclimatically, the PPNB is characterized by the high precipitation levels of the climatic optimum (Blanchet, Sanlaville, and Traboulsi 1997; Henry 1997b; Issar 2003), which suggests that climate improvement stimulated the development and expansion of farming. Many PPNB sites have been identified in the desert areas of southern

Table 6.1   Cultural phases of the Jordanian Neolithic

Phase	Time span[a]	Human-environment relations	Regional climate
Pre-Pottery Neolithic A (PPNA)	10,300–9600 BP	Scarce settlements	Dry
Pre-Pottery Neolithic (PPNB)	9600–7500 BP		Humid
Early PPNB	9600–9200 BP	Scarce settlements	Humid
Middle PPNB	9200–8500 BP	Increase of settlements, expansion of farming, and animal domestication	Humid
Late PPNB	8500–8000 BP		Dry
Pre-Pottery Neolithic C (PPNC)	8000–7500 BP		Dry
Pottery Neolithic (PN)[b]	7500–6000 BP		Humid

a Chronology based on Rollefson 2001b.

b Often referred to as Late Neolithic or Yarmoukian. The Pottery Neolithic is subdivided into PNA and PNB.

and eastern Jordan, as well as in the Negev and the Sinai. Presumably, this occupation of the drylands was the result of an ameliorated climate (Henry 1997b). In the Negev, the depletion of ^{18}O in the carbonate of land snails suggests an increase in precipitation between 9 ka and 7 ka BP (Goodfriend 1999). Similarly, ^{13}C analysis on land snail shells yields an estimate that annual precipitation in the Negev during this same period was twice the amount registered at present (Goodfriend and Magaritz 1988).

The cultural complexity of the PPNB phase resulted in a further subdivision into early, middle, and late subphases (table 6.1). The Early PPNB is poorly known in Jordan, and in some cases it is difficult to discern from the PPNA (Rollefson 2001b). As with the PPNA, the number of Early PPNB sites is small. Not until the Middle PPNB did settlements grow in number and size. The major cultural change that characterizes the Middle PPNB is the expansion of farming and the increase in domesticated species of ungulates, especially ovi-caprids, suggesting more dependence on domesticated animals than on hunting.

Floral and faunal remains recovered from Middle PPNB occupations indicate that humid conditions prevailed in areas that today are located in the deserts. Botanical macroremains from the site Jilat 7 show that its inhabitants consumed domesticated einkorn wheat, sativum-type barley, and pulses (Garrard, Baird, and Byrd 1994). Domesticated cereals

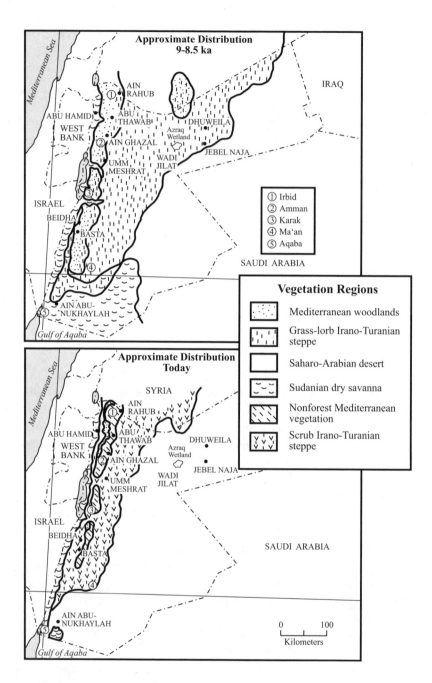

Figure 6.3 Neolithic sites mentioned in text and possible expansion of vegetation regions around 8,500 years ago as compared to present distribution.

were also found among the botanical remains recovered at the site, suggesting that, owing to improved moisture conditions during the climatic optimum, they could have been farmed locally (Henry et al. 2003). This evidence indicates that people on the steppe-desert transition were able to consume cereals although they remained heavily dependent on hunting. The ameliorated conditions may have even allowed them to farm in these areas.

Sediments from the qaʿ next to Ain Abu-Nukhaylah, a PPNB site in the Wadi Rum Nature Reserve, produced evidence of accumulated moisture. Macrobotanical remains recovered from the floors of the houses in the sites yielded remains of domesticated cereals and a series of remains of woody plants, which are now rare in the area (Henry et al. 2003). Today, this region receives 50 millimeters or less of annual rainfall. Further investigation, launched at the time of writing, is aimed to reconstruct paleoclimate using a series of proxy records, including pollen, phytoliths, diatoms, and soil development.

Ain Ghazal, the largest excavated Neolithic site in Jordan, has provided perhaps the most comprehensive record of paleoenvironmental information of all studied PPNB sites in Jordan. Located today in the transition between the Mediterranean nonforest vegetation and the Irano-Turanian steppe, the botanical remains from the Middle PPNB deposits at Ain Ghazal suggest a more wooded environment, a premise that is supported by the abundance of oak wood used in the construction of dwellings and the presence of a set of fauna typical of oak woodland (Köhler-Rollefson and Rollefson 1990; Köhler-Rollefson, Quintero, and Rollefson 1993). During this time, the consumption of meat of domesticated animals increased in the region around Ain Ghazal, apparently because the deterioration of the environment may have reduced the habitats of some of the formerly hunted species (Köhler-Rollefson 1997). This change is also evident in the faunal remains of many other sites of the Late PPNB.

The Middle PPNB occupations of Basta, similarly located on the ecotone of nonforest Mediterranean vegetation of the southern highlands and the Irano-Turanian steppe, provide a scenario that parallels the cultural ecological events in Ain Ghazal: As bones of typical woodland game such as deer and boar decreased, bones of domesticated sheep and goats increased. As time passed, not only did hunting decrease in favor of consumption of domesticated animals, but also the number of typical game animals of the steppe such as gazelles and horses decreased as domesticated animal populations increased (Rollefson 2001a). This gradual change of balance may be interpreted as a result of depleted woodlands in favor of steppes or as an indication that the radius of hunting around the site grew to include the steppe area.

The end of the Middle PPNB is characterized by several cultural changes often attributed to human transformation of the environment. The most evident change is the decline of farming and the increase of pastoral nomadism (Rollefson 2001a). The explanation for that change usually alludes to either land degradation or seismic activity, which is consistent with the information obtained from excavations (Rollefson 2001b). Climatic deterioration is another possible cause of societal-environmental deterioration at the end of the Middle PPNB. However, the low resolution of paleoclimatic proxy data complicates the pinpointing of short-term dry phases. Pollen from Lake Hula shows high arboreal pollen frequencies during the period 9–8 ka BP, although some peaks and lows in the curve are noticeable (see fig. 5.2). Yet even the lows in arboreal pollen during this time are relatively high compared to the rest of the Holocene. Unfortunately, determining the time length represented by such peaks is difficult. If we assume that they encompass 100- to 200-year intervals, the resolution is not enough to permit us to assess the duration of short-term dry events that could have had negative consequences in agricultural development and small village economy. In this case, for example, a 20-year drought period could be long enough to derail agricultural development.

Not until the end of the Late PPNB did climatic conditions deteriorate (Issar 2003). Proxy data, both local and regional, suggest that both human and climatic causes were combined in the environmental deterioration at the end of the Middle PPNB. The evidence of agricultural decline and landscape deterioration is evident in most records, suggesting that this was the first large-scale environmental crisis since farming and pastoralism were implemented.

The Late PPNB is characterized by a transition from a mixed economy (farming and pastoralism) to a high dependence on nomadic pastoralism and hunting (Rollefson 2001a). The Late PPNB saw the emergence of the so-called megasites (usually larger than 10 hectares) such as Ain Ghazal, Wadi Shuʿeib, and Basta (Rollefson 1998). These megasites, or aggregated villages, seem to have collapsed by 8900 cal. BP, probably as a result of severe overexploitation and a decrease in precipitation (Rollefson 2001b). Overall, the record indicates the disappearance of large aggregated villages is followed by the emergence of small-scale farmsteads (Kuijt 2000). Ecological and settlement changes during the PPNC can be seen, for example, at Dhraʿ (Kuijt 2000) and Wadi Shuʿeib (Simmons et al. 2001). The small sites prevailed through the PPNC and Pottery Neolithic. Not until the Chalcolithic, about 1,500 years later, did the large aggregated settlements reappear (Kuijt 2000).

Compared with those of the PPNB, faunal remains of PPNC occupations contain more domesticated ungulates than wild species, which is a trend that continues into the ceramic Neolithic (Simmons et al. 1988). Climatic conditions and environmental quality during the PPNC were most likely similar to those of the Late PPNB, that is to say, drier and with scarce resources (Rollefson 1998). This suggests a breach in what was nomadic pastoralism and sedentism, an aspect that became clear in subsequent cultural periods. Overall, because of extreme land-use changes and overexploitation of resources amid shifting climate, the Late PPNB and the PPNC may be considered a major, and by far the first, ecological crisis of the Holocene period in the Levant.

The Pottery Neolithic (PN) is poorly known in Jordan because it has received little attention since most Neolithic research has focused on the origin of agriculture (Kafafi 1992). In Jordan, the most common ceramic traditions of this phase are the Yarmoukian, Jericho, and Wadi Rabah (Kafafi 1998). The best-known sites for this period are Ain Ghazal, Abu Thawab, Wadi Shuʿeib, Ain Rahub, and Abu Hamid. However, dozens of other sizable sites have been identified in surveys but not studied in detail. Although apparently most abundant in the woodlands, some sites of this period have also been reported in the drylands (steppe and deserts), where activity seems to have been oriented toward pastoralism and hunting (Rollefson 2001b).

A relevant characteristic of settlement in the Pottery Neolithic is that some of the sites abandoned during the Late PPNB and the PPNC were resettled (Banning 2001). This could be interpreted as improvement in environmental conditions or perhaps a technological change that once again allowed more stable settlements. The rapid adoption of pottery is hypothesized to have facilitated and secured grain storage, but most of the pottery of this period seems not to have served such a purpose (Banning 2001). Climatic amelioration seems to have played a role in moving populations back to a more sedentary life, focusing on the redevelopment of mixed agriculture. However, the emphasis on farming and sedentary life during the PN led to the segregation of nomadic pastoralism, hence creating the breach between nomadic shepherds and settled farmers, a dichotomy that would prevail for millennia.

Climatic records indeed show an increase in rainfall, although a few sets of proxy data suggest contradictory developments. For example, lake-level changes in the Dead Sea do not support climatic improvement that otherwise is evident in stable isotopic data from Soreq Cave (fig. 6.4). Although the Pottery Neolithic phase is known to have enjoyed a mostly wet climate, dry spells are evident in some records such as pollen (Baruch and Bottema 1999) and stable isotopes (Bar-Matthews, Ayalon, and Kaufman 1998). The PN is the last cultural phase to experience the final benefits of the climatic optimum of the 9–6 ka BP period. The transition from wetter to drier climates gradually occurring during this time may have had an ecological impact on the environment, particularly in the hydrological regime of streams and in vegetation cover.

Reduced rainfall may have had an impact on the recharge of alluvial aquifers, which consequently resulted in a lower water table in floodplains. Under such conditions, stream incision was more likely to occur. In Wadi ath-Thamad, water table lowering was probably one of the causes for the downcutting of the Thamad alluvium and the eventual destruction of the Umm Meshrat 1 site by fluvial erosion (Cordova et al. 2005). Site damage by channel erosion and rapid accumulation of sediment from upstream erosion have been reported in other alluvial locales such as Wadi Faynan and Wadi Ghuwayr (Barker et al. 1997, 1998), Abu Hamid (Dollfus et al. 1988), and Beidha (Field 1989), as well as in numerous sites in the Negev and the Sinai (Goldberg 1986, 1994). For around the same period as the examples above, evidence of incision and channel erosion with subsequent accumulation of sediments has been reported along the streams of the East Bank of the Jordan (Mabry 1992).

This unstable geomorphological landscape prevailed through the early part of the Chalcolithic period. These intense geomorphological changes in wadis indicate that some PN sites may have been severely transformed or even destroyed, which needs to be taken into consideration when assessing settlement patterns in alluvial environments.

Human Impacts on the Landscape

Increasing population, farming and pastoral activities, and settlement expansion during the Neolithic phases had an impact on the landscape. Evidence for such an impact exists in the records of floral and faunal

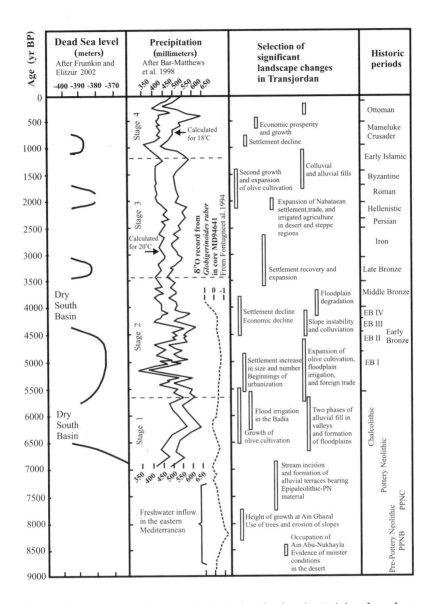

Figure 6.4 Selection of geomorphological and palynological data from four localities in the context of paleorainfall (based on δ¹⁸O data) and Dead Sea–level fluctuations during the Holocene, correlated with landscape change and cultural developments.

remains recovered from cultural and natural deposits. In general, these records show deterioration of woodlands, reduction of animal species through hunting, and increasing emphasis on domesticated sheep and goats (Köller-Rollefson and Rollefson 1990).

The human transformation of vegetation at the end of the Neolithic period reached levels beyond the threshold of resilience during preagricultural times. Accordingly, the impact of farming and grazing led to the formation of anthropic vegetation types such as the nonforest Mediterranean vegetation and the scrubby vegetation that today characterizes the Irano-Turanian steppe most likely began to develop (fig. 6.1).

Relatively abundant rainfall in conjunction with human disturbance may have had a negative impact on the environment by contributing to the rapid erosion of soils, hence producing sediments that accumulated in valley bottoms and streams, eventually destabilizing floodplains. In some cases, Natufian and older occupations were buried under alluvium and colluvium, as happened at Beidha (Field 1989) and Umm Meshrat 1 (Cordova et al. 2005). Sequences of poorly developed A soil horizons in vertical sequences of alluvial deposits suggest that alluvial accumulation did not occur all at once but in successions combining relatively stable short periods (A-horizon formation) followed by rapid accumulations of alluvial silts and gravel.

Soil erosion and rapid alluvial and colluvial accumulation are evident on the slopes. At Ain Ghazal, persistent accumulation of red sediments has been identified throughout the early phases of occupation of the site (Mandel and Simmons 1988), indicating periodical mobilization of hillslope sediments. Below the accumulations of colluvium, older soils lack A horizons, suggesting that topsoil removal was intense during this period (Mandel and Simmons 1988; Simmons et al. 1988).

Paleobotanical, paleozoological, and geomorphological data associated with Neolithic sites, as well as regional paleoecological records, have yielded evidence that the first human-led transformation of the landscape had a tremendous impact not only on the Neolithic communities but also on the later landscapes that were inherited by the peoples of subsequent periods. The Neolithic merely witnessed the introduction of new forms of subsistence. Later cultures, however, took some of the new subsistence forms to greater levels. Agriculture, for example, underwent intensification, especially through irrigation and commercialization. The result was an entirely new phase in the history of millennial landscape change.

The Middle Holocene and the Rise of Urbanism

Cultural Development and Climatic Context

The Middle Holocene corresponds roughly to the Chalcolithic and Early Bronze Age. Like the Early Holocene, it is a period of intense landscape transformation in large part attributable to intensification of human activities. Numerous remains of cultivated plants in sites of the arid parts of the Rift Valley (e.g., Teleilat el-Ghassul and Tell el-Hayyat) suggest prosperous communities, which presumably enjoyed an ameliorated climate. Regional paleoclimatic records show a relative increase in precipitation, interrupted by relatively short dry episodes at 6.5 ka, 5.2 ka, and 4.2 ka BP (Weiss 2000; Issar 2003).

Since the early days of archaeological research, scholars have observed that most Chalcolithic and Early Bronze Age sites were located in the immediate proximity of streams (Albright 1925). For this reason, most archaeologists have associated sites of these periods with floodplain irrigation, despite a lack of clear evidence of irrigation structures except in the wadi next to the site of Jawa on the Northwestern Basaltic Plateau. In the area around the site, dams and diversion walls have been dated to 3000 BC at Jawa (Helms 1981; Betts 1991), which is the oldest evidence of early irrigation in Jordan.

Paleobotanical evidence obtained from a few localities in the Levant suggests that crops were watered during the Chalcolithic and Early Bronze Age. In one study, the distinctive morphological characteristics of silica cells (phytoliths) of irrigated cereals were recovered from Chalcolithic cultural deposits in southern Israel (Rosen and Wiener 1994). In Wadi al-Wala, the pollen assemblage from the Iskanderite Soil, dated to the Chalcolithic and Early Bronze Age, shows high amounts of cereals in association with relatively high amounts of aquatics such as *Typha* sp. (cattail), *Butomus* sp. (flowering rush), and *Lemna* sp. (duckweed) (fig. 5.4-A). This evidence suggests two possible scenarios: that this soil was irrigated with stagnated water (kept perhaps behind a dam), or that the floodplain was drained for agriculture. Archaeobotanical records from sites in the Jordan Valley (Mabry 1992; Fall, Lines, and Falconer 1998) provide abundant evidence of cereal production, even though this region is arid. Today, this area receives less than 100 millimeters of annual rainfall. This is suggestive that somehow irrigation was practiced. But where are those irrigation structures?

The lack of remains of irrigation installations dated to this period may be attributable to three possible factors. One is that the irrigation systems practiced in most localities involved a simple form of floodwater diversion for which permanent stone structures were not necessary. The other possible explanation for the absence of irrigation diversion structures is that they were obliterated by the same incision and erosion processes that eventually destroyed the floodplain at the end of the Early Bronze Age (see below). However, it is not possible to generalize an irrigation system for a region as diverse as Transjordan. Most likely, irrigation systems in the wadis in the Jordan Valley differed from those on the plateau, but not until a study focuses on the wadis will details on water management for the Chalcolithic and Early Bronze Age become known.

Regional settlement patterns during the Chalcolithic and Early Bronze Age are notably different from those of the previous and later periods. Regional differences in Chalcolithic sites (fig. 6.5-A) have been distinguished (Bourke 2001). The recurrent settlement-near-floodplain pattern was very common in the drylands of the Jordan Valley (region I) and the Plateaus (region II) (Harrison 1997). However, exceptions exist, because location patterns differed in relation to their economic specialization. The situation was completely different on the Ras an-Naqb Plateau in southern Jordan (region III), where settlements were less permanent and linked to movements across different environments (Henry 1995). This alludes to a strong dependence on husbandry or perhaps to strategies linked to soil and water procurement for agriculture. Settlement patterns in the area between Wadi al-Karak and the northern end of Wadi Araba (region IV) were similar to those of region I. The area of the Rift Valley between Wadi al-Hasa and Wadi Faynan (region IV) had a unique pattern. Most sites were concentrated around copper mining activities, although agricultural and pastoral economies were an important part of the regional economy as well (Barker 2000). The area between Wadi Zarqa' and Wadi Nimrin (region V) is devoid of large Chalcolithic settlements, probably because no major streams flow through it (Bourke 2001). Outside the regions mentioned above, very few Chalcolithic sites have been reported.

Unfortunately, compared with other areas of the Levant (e.g., Palestine), there are few excavated Chalcolithic sites in Transjordan. The best known of the excavated sites in Jordan is Tuleilat al-Ghassul, which is one of the most frequently referenced sites in the literature of the Chalcolithic period of the Levant (Bourke 2001). The scarcity of excavated sites

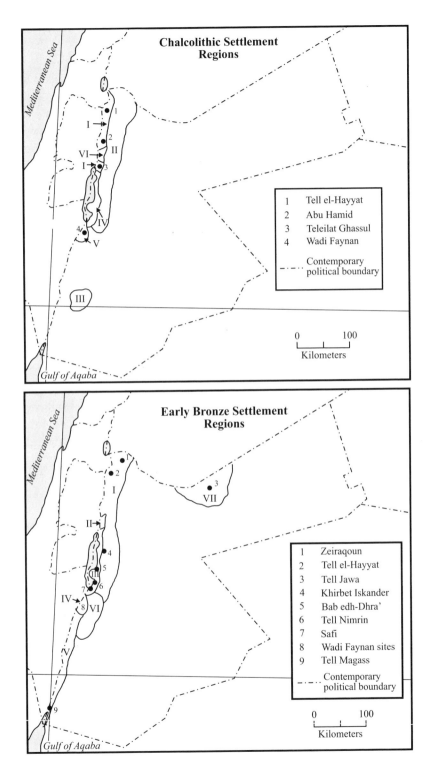

Figure 6.5 Chalcolithic and Early Bronze sites and regions mentioned in text.

also results in a scarcity of paleoenvironmental information, especially for regions outside the Jordan Valley.

Chalcolithic villages of the Levantine region were generally large and complex, similar in size to those of the Middle PPNB. The use of new technologies, such as the use of copper and the practice of irrigation, suggests cultural diffusion through migration and contact with peoples outside the Levantine region (Bourke 2001). Chalcolithic societies coped not only with drier climates and a more deteriorated environment but also with a more changing climate than their Neolithic predecessors experienced. Stable isotope records for the Chalcolithic and Early Bronze Age show considerable ups and downs in precipitation (fig. 6.4). This alternation between exceptionally wet and dry climates marks the transition between the humid conditions of the Early Holocene (i.e., the climatic optimum) and the drier conditions of the Middle and Late Holocene (Bar-Matthews, Ayalon, and Kaufman 1998). Nonetheless, social and technological innovations, whether indigenous or introduced, led to the expansion of settlements that culminated with the creation of proto-urban centers during the Early Bronze I (EB I) period.

The division of the Early Bronze Age into EB I, II, III, and IV is largely based on variations of pottery types (table 6.2). This sequence also corresponds to significant changes in settlement size and economic development that shows the development of large villages into towns. Although towns experienced continuous growth during the EB I, EB II, and EB III phases, sudden urban decline and increased nomadism characterize the EB IV. An important socioeconomic characteristic of the EB I–EB III time is the increasing intensification of regional and foreign trade relations (Philip 2001). Thus, Palestine and Transjordan expanded economic relations with the Nile Valley and Mesopotamian empires (Butzer 1997).

Rural economies emphasized a typical Mediterranean mixed economy based in part on the cultivation of olive, grapes, figs, and cereals and in part on nomadic husbandry (Fall, Lines, and Falconer 1998). Pollen records indicate a high increase in *Olea* pollen frequencies, reflecting the widespread cultivation of olive (Baruch 1990, 1993; Schwab et al. 2004). This is also supported by the abundance of *Olea* pollen in local pollen sequences of mid-Holocene soils in Wadi Shallalah and Wadi al-Wala (see fig. 5.4) and the large quantities of olive seeds in the archaeobotanical records recovered from Chalcolithic and EB cultural deposits (Neef 1990, 1997).

Livestock herding remained an important activity during the Early

Table 6.2   Early Bronze Age subdivisions and socioeconomic characteristics

Phase	Time ranges (absolute dates BC)	Regional climate
EB I	3600–3100/3000	Moister than today,
Early EB I	3600–3400/3330	with short drought
Late EB I	3400/3300–3100/3000	episodes
EB II	3100/3000–2850/2750	Moister than today but highly variable
EB III	2850/2750–2400/2300	Moister than today but with a trend toward aridization
EB IV	2400/2300–2000	Like today or drier

Sources: Time ranges for EB I–EB III are based on Philip 2001 and EB IV on Palumbo 2001.

Bronze Age, which supplemented cultivated foods in times of crisis (Fall, Lines, and Falconer 1998). The combination of farming based on the production of cereals, olives, grapes, and other orchard products with livestock husbandry formed the basis of the rural economies around the EB settlements. During this time, a type of pan-Mediterranean agricultural economy developed in Transjordan, as it did in other parts of the Levant (Stager 1985).

The settlement geography of the Early Bronze Age continued the regional differentiation attained during the Chalcolithic period (fig. 6.5). Most EB towns grew out of preexisting Chalcolithic villages (Harrison 1997; Philip 2001). Most of the large EB sites are located in the western part of Jordan, predominantly along the Jordan Valley and on the plateaus from the Yarmouk River to Wadi al-Hasa (region I) (fig. 6.5 bottom). Sites are almost absent between the Zarqaʾ River and Wadi Nimrin (region II), as was the case during the Chalcolithic. Sites in the area of the Rift Valley between Wadi Nimrin and the southern Ghor (region III) are very sparsely distributed and dominated by three relatively large settlements — Bab edh-Dhraʿ, Nimrin, and Safi.

The Wadi Faynan area (region IV) had concentrations of settlements around the copper mines. The area between Wadi Faynan and Aqaba (region V) had very scarce small settlements, although a large settlement (Tell Magass) stood in the far south of this region. The highland region south of Wadi al-Hasa (region VI) was also characterized by small and scattered settlements.

Despite the ameliorated climatic conditions, the deserts were very sparsely populated during the Early Bronze Age. However, settlements were slightly more abundant on the Northwestern Basaltic Plateau (region VII) than elsewhere in the drylands. Irrigation-dependent settlements in region VII show parallel developments in the west. As mentioned above, the irrigation system associated with Jawa is one of the examples of advanced water management technology for this time.

Although most EB settlements were concentrated in the western part of Transjordan, they did not necessarily occupy humid regions of the highlands. Numerous large settlements were located in the dry regions of the Rift Valley (as can be appreciated on the EB settlement maps by Savage [2005]). This pattern reflects the preference for areas with potential for flood irrigation over rain-fed agriculture and proximity to the main trade routes. As in the Chalcolithic, the area between Wadi Zarqa' and Wadi Nimrin (region IV) remained largely unoccupied during the Early Bronze Age. This may be attributable to a lack of drainages with potential for irrigation and fertile floodplain soil (Mabry 1992).

Although climate during the Chalcolithic and the EB I, II, and III sub-periods was generally more humid than it is today, short-term droughts occurred, as shown by $\delta^{18}O$ from Soreq Cave (fig. 6.4). High atmospheric and water temperatures played an important part in enhancing cyclogenesis in the eastern Mediterranean Sea. Alluvial sediments in Nahal Lachish and other wadis in the Shephelah region of Israel present evidence of moist climatic conditions (Rosen 1986, 1995). Soils, sediments, and pollen in Wadi al-Wala and Wadi Shallalah also point to a rather humid climate (Cordova 2005b). The rapid and unstable fluvial conditions that characterized the Late Neolithic and part of the Chalcolithic seem to have reached stability by 5 ka BP. This is verified by organic soils that developed in floodplain deposits dated to this time in the streams of the Jordan Valley (Mabry 1992) and the Western Highlands (Cordova et al. 2005).

Human Impacts on the Landscape

Historical records provide indirect information on the effects of urban and rural economies on the environment of the Near East during the Early Bronze Age. Egyptian and Mesopotamian sources, for example, testify to the consumption of Lebanese cedar (*Cedrus libani*), which eventually led to the depletion of cedar forests in the Lebanese Mountains (Rowton 1967; Redford 1992). However, this depletion of cedars is not clear in the pollen records of the Ghab Valley in Syria, which is located on the east side of the mountains. The pollen record fails to detect the reduction of cedars during the Early Bronze Age because cedar pollen frequencies were already low; the rapid decline occurred much earlier, during the Pre-Pottery Neolithic, presumably because of the effects of rapid expansion of early agricultural communities (Yasuda, Kitagawa, and Nakagawa 2000).

Written records from Egypt also refer to large imports of olive oil from Levantine cities during this period (Redford 1992). The pollen diagrams from the lakes of the Rift Valley show an increase in olive cultivation (see fig. 5.2). The pollen diagrams also illustrate that the areas formerly covered with oak forests were being replaced with cultivated olive (Baruch 1990; Schwab et al. 2004). Overall, the historical sources account for substantial trade of agricultural products with Egypt and other regions, which in turn implies intensification of rural activities.

Transformation of herbaceous vegetation may have occurred through the increase of grazing as growing town populations required meat, milk, and other livestock products. However, the initial impacts of goat and sheep grazing during the Neolithic and Chalcolithic periods had already transformed vegetation tremendously. Pollen diagrams from the Ghab Valley show high pollen frequencies of taxa linked to anthropicized herb and scrub (e.g., Cruciferae and *Sanguisorba*-type) during the periods preceding the Early Bronze Age. This contradicts the idea that Chalcolithic peoples dealt with a forested and somehow unspoiled landscape, as is often implied in some human ecological models (e.g., Gophna, Liphschitz, and Lev-Yadun 1986).

Research on copper mining and smelting in the Wadi Faynan region of southern Jordan has provided evidence of the impact that these activities had on the landscape (Levy, Adams, and Shafiq 1999; Barker 2000). The depletion of wood in the forests of the escarpment west of Wadi Faynan is attributed to the production of charcoal used for smelting (Engel 1993). In addition, smelting increased air pollution to levels comparable to those of modern industrial regions (Hauptmann and Weisberger 1987; Barker 2000).

Floodplain degradation is evident in some valleys of the Levant at the end of the Early Bronze Age (see fig. 5.7). However, whether floodplain degradation was the result of climate change or human mismanagement of the basins is unclear (Mabry 1992).

The End of the Early Bronze Age: An Environmental Crisis?

The ancient civilizations in the Near East reached their zenith during the first centuries of the third millennium BC. But at the end of the millennium, sudden collapse struck the Old Kingdom civilization of Egypt; the Akkadian Empire of Mesopotamia; the Bronze Age civilizations of Syria, Palestine, Greece, and Crete; and civilizations located as far distant as India (Weiss 2000). This widespread event of civilization decline affected societies in Transjordan, where settlement size reduction and abandonment are evident in the archaeological record (Richard 1980; Donahue 1981, 1984, 1988; Prag 1984; Mabry 1992; Palumbo 2001; Philip 2001).

Was the collapse of Near Eastern civilizations at the end of the third millennium BC the result of climatic change or socioeconomic troubles? This is the question that haunts archaeologists, historians, geographers, and other scholars studying environment and society in the ancient Old World. In pursuit of an answer to this question, multidisciplinary groups of scholars in the late 1980s and early 1990s began to combine their efforts. In particular, the compilation of papers dealing with this question edited by Nuzhet Dalfes and colleagues (1997) stirred further interest in the subject. Nevertheless, the variety of data and interpretations has not produced a final consensus.

The complexity and geographic variation resulted in a series of possible variables in the process of societal collapse, which inevitably led scholars to seek either climatic or social causes for the crisis. Those in favor of climatic causes point to a series of climatic fluctuations recorded in various parts of the world. Harvey Weiss (2000) examined paleoclimatic records at local, regional, and global scales to conclude that a 300–400-year period of low precipitation took place in several regions of the world. Evidence of intense dust storms exists in sediment cores from the Gulf of Oman, where an increase in quartz and other minerals indicate the presence of dust particles deposited in the sea (Cullen et al. 2000). The investigators of this study were able to trace the origin of this dust to Upper Mesopotamia. Subsequently, the age of dust particles lying at the bottom of the Gulf of Oman was associated with stratigraphic data at Tell Leilan in Upper Mesopotamia, showing that the dust storms are correlated with the collapse of civilization in that region. Beyond the Near East, deep-sea and ice cores reveal changes in atmospheric circulation that can be tied to diminished monsoon activity in the Indian Ocean.

Arie Issar (2003) attributes salinization in the Mesopotamian low-lands to climatic deterioration, which contradicts the previously pro-posed idea that salinization was caused by irrigation mismanagement (Jacobsen and Adams 1958; Artzy and Hillel 1988). Issar (2003) concludes that regardless of the negative effects of irrigation practices, soil salin-ization in the large floodplain soils was the result of climatic desicca-tion, which is evident in numerous records of oxygen and carbon iso-topes. Miryam Bar-Matthews, Avner Ayalon, and Aaron Kaufman (1998) present high-resolution data for the reconstruction of paleoprecipita-tion based on oxygen isotopes from Soreq Cave, which show a dry phase around 4 ka BP (fig. 6.4).

The only high-resolution proxy records that detect short-term cli-matic deterioration in the Levant at the end of the third millennium BC are the isotopic data from the Sea of Galilee (Stiller et al. 1983–84), the speleothems in Soreq Cave (Bar-Matthews et al. 1998, 1999), and the sea-floor sediments of the eastern Mediterranean Sea (Fontugne et al. 1994). Geomorphological and botanical records from salt caves on the south-western shore of the Dead Sea show a lake-level drop that may be related to climatic desiccation (Frumkin et al. 1994; Frumkin and Elitzur 2002).

Pollen records from lake sediments in Turkey, Syria, and Iran show evidence of environmental deterioration, but this was not necessarily the result of a change to a drier climate (Bottema 1997; Butzer 1997). The low sedimentation rates in Near Eastern lakes preclude analysis of pos-sible vegetational responses to a relatively short dry period (e.g., 300–400 years). Pollen records from lakes in the Rift Valley and the Golan Heights do not show vegetation change directly associated with climatic deterioration around 4,000 years ago (see fig. 5.2). Instead, they show a decline in *Olea* pollen around this time, suggesting a widespread aban-donment of olive cultivation that is mirrored in local pollen records for the same period in Wadi Shallalah and Wadi al-Wala (see fig. 5.4).

Although substantial evidence for climatic deterioration at the end of the third millennium BC exists, problems with resolution at the local scale still cloud the magnitude of this climatic event. Consequently, criti-cism of the climatic hypothesis arose among some scholars, leading to the upsurge of nonclimatic hypotheses. Those in favor of nonclimatic causes provide examples of socioeconomic crises that may have occurred regardless of climatic deterioration (Butzer 1997). On the basis of historical and paleoenvironmental data from various regions of the Near East and the eastern Mediterranean, Karl Butzer (1997) discusses a possible scenario in which economic crises in one region were transmitted to neighboring regions, creating a "domino-fashion" collapse of socioeconomic systems. He explains his argument using the collapse of the Old Kingdom in Egypt, which impacted the network of Egyptian trade points across the Levant and surrounding regions. From here the crisis was transmitted to neighboring areas dependent on resources and trade with the collapsed regions. In this context, drought and accumulated stress on the land could have been factors involved in the collapse of economic systems, but they would not have been decisive ones. Although Butzer's proposed scenario stresses the economic interdependency between regions, it does not conflict with the idea of climatic deterioration as a contributory factor to the collapse of urban societies.

As an alternative to purely climatic or purely socioeconomic causes, Tony Wilkinson (1994, 1997) presents a comprehensive model to illustrate the development and collapse of Early Bronze sites in the Khabur Plains of Upper Mesopotamia. This region is located on the Irano-Turanian steppe. Today the Khabur Plains receive an average of 200 millimeters of rainfall a year, presenting the opportunity for rain-fed agriculture, but with high variability that makes agriculture a risky activity. In addition, the main streams, which are tributaries of the Euphrates, present irrigation potential as long as flow is constant. In many aspects, this region is similar to the Irano-Turanian region of Jordan, that is to say, relatively low and highly variable annual rainfall, highly erodible thin soils, and streams with potential for flooding irrigation.

The foundations of Wilkinson's model for explaining the settlement collapse in the Khabur Plains lie in the capacity of these settlements to use the natural resources around them while implementing strategies to procure food and other necessities in a highly variable environment. This model states that as the population grows, the vulnerability of resource procurement systems increases. Consequently, the growing stress of an increasing population pushes the environmental and social systems to the brink of failure and eventually results in socioeconomic collapse and abandonment of sedentary life. The model considers the interplay of several climatic and economic factors and implies that societies contain their growth to avoid crossing the threshold of their carrying capacity.

However, this threshold dropped as climate in the Near East deteriorated, thus increasing the probability of economic collapse.

Western Transjordan at the End of the Third Millennium BC

The fourth and third millennia BC in western Transjordan were characterized by rapid urbanization, intensification of olive, grape, and grain cultivation, and an increase in interregional interaction between different regions of Transjordan and abroad (Fall, Lines, and Falconer 1998). This phase was followed by a rapid process of urban and economic decline between 2200 and 2000 BC, which corresponds with the EB IV phase (Palumbo 2001; Philip 2001). Despite decline and abandonment, a few settlements remained occupied through the EB IV, as is the case of Khirbet Iskander (Richard 2000; Palumbo 2001).

Khirbet Iskander is used here as an example through which to evaluate the possible scenario of rise and decline of settlement in the Middle Holocene. This scenario draws on geoarchaeological data obtained from the alluvial deposits of Wadi al-Wala and archaeological data obtained through excavation (Richard and Long 1995; Richard 2000; Richard, Long, and Libby 2001).

The main focus of geoarchaeological research is on the Iskanderite Alluvial Inset, which is the erosional remnant of the floodplain that existed in Wadi al-Wala during the Chalcolithic and Early Bronze Age (see fig. 5.8). Paleoecological data consist of pollen data obtained from the Iskanderite Soil and the underlying Iskanderite Alluvial Inset (see fig. 5.4 top). The Iskanderite Soil yielded a pollen assemblage with a mixture of cultivated plants and weeds. Very few native trees (oak and pistachio) seem to be present in the area at this time, but the amounts of their pollen are higher than in the present. The colluvial deposit (Terrace II colluvium) mantling this soil was deposited before stream incision eroded the floodplain. This colluvium represents the deposition of sediments washed from the slopes, which seem to have been terraced. Remains of such terraces are still found on the slopes. Whether this colluvium was deposited because of negligence of the terraces or because of a sudden erosional event linked to climate change is not clear. The presence of aquatic plant pollen in this colluvial deposit indicates that the floodplain was still wet when the colluvium was accumulated. Soon afterwards, the floodplain was cut down by the force of water, presumably because of a lower water table.

The sequence of change comprising abandonment, colluvial deposition, water table lowering, and incision could have occurred over a period of decades. In this case, site abandonment could have occurred gradually, perhaps encompassing the entire two to three centuries of the EB IV phase. Accordingly, the sequence of cultural and ecological changes leading to the rise and decline of Khirbet Iskander between 4000 and 2000 BC can be summarized in the context of population growth, carrying capacity, agricultural development, urban growth, extraregional trade, climatic changes, and floodplain dynamics (fig. 6.6).

Adverse environmental conditions that converged at the end of the third millennium BC (i.e., climatic deterioration and intense use of land during boom periods) may have reduced the carrying capacity of the landscape, which consequently would have had a negative impact on the economies of EB III towns (Cordova 2005a). Furthermore, population decline may have had a negative impact on the maintenance of terraced olive plantations on steep slopes. Thus, prolonged droughts followed by flash-flood runoff events washed down sediments from abandoned terraces, leading to rapid downslope mobilization of colluvial sediments and their subsequent deposition on the edge of the floodplain (fig. 6.6). Following agricultural terrace abandonment, intense flash floods and the lowering water table led to incision and destruction of the floodplain that had once provided settlers with the means for irrigated agriculture. Perhaps Khirbet Iskander was abandoned at the end of EB IV and was never again resettled because the fertile floodplain was no longer there and soils on the slopes were eroded.

Floodplain incision and erosion seems to have been widespread in the Levant and other areas of the Near East at the end of the Early Bronze Age or shortly afterwards (see fig. 5.7). These events occurred always in the vicinity of Early Bronze Age sites that underwent decline at the end of the third millennium BC (Donahue 1981, 1984, 1985; Rosen 1986, 1995, 1997b; Mabry 1992; Wilkinson 1997, 1998, 1999). Therefore, the hypothesis originally established by William Foxwell Albright (1925) regarding the importance of wadi flood irrigation in the success of EB towns could be a fact. Nonetheless, more detailed geoarchaeological data around other EB sites may improve the regional picture of the environmental crisis that occurred at the end of the third millennium BC.

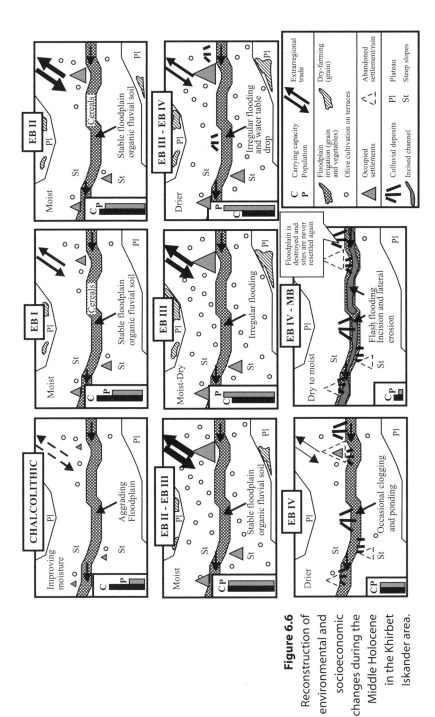

Figure 6.6
Reconstruction of environmental and socioeconomic changes during the Middle Holocene in the Khirbet Iskander area.

The Late Holocene and the Formation of Modern Landscapes

Cultural Changes in Their Climatic Context

Landscape changes during the past four millennia are better known than in earlier times because of the larger amount of proxy data and historical information. This time period encompasses stages 3 and 4 in the curve of precipitation reconstructed from $\delta^{18}O$ (fig. 6.4). Stage 3 represents precipitation levels lower than those of the Early and Middle Holocene, but with minor variations. Stage 4 presents two peaks of precipitation and a low in between them. The first peak seems to occur at the turn of the first millennium AD and is probably associated with the Medieval Warming. This event roughly coincides with the Crusader and Mameluke historical periods. A low in precipitation occurs around 500 years BP. The subsequent four centuries, roughly corresponding to the Ottoman period, is characterized by an increase in precipitation, which peaked sometime between 300 and 200 years ago. Since then, precipitation levels have declined, particularly during the nineteenth century.

Human Impacts on the Landscape

Although climatic fluctuation undoubtedly had an impact on the landscape, human-led transformation is considerable, particularly because during this period, agriculture and irrigation were implemented in many regions. In general, changes associated with human activities and climate are better expressed in changes in vegetation, soils, and streams.

Of all the pollen sequences of the southern Levant, the one that provides the best resolution for the Late Holocene is the core from Birkat Ram (see fig. 5.2). Some interesting data are also provided by sequences obtained from the Dead Sea (Heim et al. 1997). The other sequences (Hula Lake, Lake Kinneret, DS-I, DS-II, and DS-III) show recurrent changes, despite the low resolution. Oaks and pistachio, the two native trees, are in general low during these four millennia. The most significant change occurs again in the curve of olive pollen, which increases following the Hellenistic period and decreases at the end of the Byzantine period, during which time olive was extensively cultivated.

Pollen records show that vegetation change was less intense during the Late Holocene than the Early and Middle Holocene. During the early and middle parts of the Holocene, intense vegetation changes related to the initial deforestation for early farming and the initial effects of pastoralism. The land was not adapted to these new activities, as it is today. The pollen diagrams clearly illustrate that the most dramatic changes occurred between the beginning of the Neolithic and the end of the Early Bronze Age, that is to say, from the beginnings of agriculture through the first phase of urbanization.

The widespread destruction of vegetation alleged to have occurred during the early parts of the Islamic period and afterwards as a result of the increase of pastoral nomadism (Lowdermilk 1944; Reifenberg 1955) is not recorded in the pollen diagrams. Indeed, the notion of nomadization and vegetation destruction after the Islamic conquest is not supported by the archaeological and paleobotanical record (Kedar 1985).

The alleged destruction of vegetation during the Ottoman period, however, is well documented in historical records, particularly through the acquisition of timber for the construction of the Hejaz Railway (Pick 1990). Additional data include the exploitation of timber in the Western Highlands by Circassians who traded it in the markets of Jerusalem (Pick 1990). The deterioration of the Transjordanian landscape at the end of the Ottoman period is clearly portrayed in the accounts of travelers (e.g., Huntington 1911) and in the accounts of events and bureaucratic changes (recorded and commented on in Lewis 1987 and Abujaber 1989). Another aspect that shows a strong relationship with landscape degradation, and in particular with soil erosion, is settlement abandonment, as shown in a study of environmental implications of settlement change in the Wadi al-Hasa region (Hill 2002).

Intense soil erosion seems to have occurred mostly between the time of early village farming and the formation of complex societies in the Early Bronze Age, that is to say, between 9 ka and 5 ka BP. Soils and sediments around Neolithic sites show the intensity of soil erosion during this time. Examples can be found in Ain Ghazal (Mandel and Simmons 1988), Beidha (Field 1989), Abu Hamid (Dollfus et al. 1988), and Umm Meshrat (Cordova et al. 2005), among others. Large amounts of sediments deposited at the foot of Chalcolithic and Early Bronze Age sites and valleys nearby attest to the continuation of intense erosion. The amount of sediment deposited during this period in the wadis of the eastern bank of the Jordan Valley (Mabry 1992), which is correlative with the Tur al-Abyad alluvial fill in Wadi ath-Thamad (Cordova et al. 2005), is evidence of intense sediment deposition in the streams of western Jordan.

Many of the Early Bronze Age sites of the plateaus of the Western Highlands stand on red soil, while Iron Age sites stand on bedrock. This pattern is evident, for instance, in Madaba, where the EB foundations rest on *terra rossa* (i.e., red Mediterranean soils) (Harrison et al. 2000) as opposed to older parts of the tell, which are simply standing on limestone. The Chalcolithic-EB site Zeiraqoun on the Irbid Plateau (see fig. 5.7) rests on red Mediterranean soils, while the area around the tell is a barren surface. This suggests that the tell itself protected the soil underneath, while the lands around it were eroded. A similar pattern is seen also in the EB sites of the Wadi al-Wala area (e.g., Khirbet Iskander, Abu Khirqa), which stand on red soil, while the Iron Age and later sites lie on rock. Overall, these examples show that by the beginning of the Iron Age, not much soil existed on the plateaus, and the eroded limestone landscapes of the plateaus looked very much like the present ones.

Alluvial and colluvial deposits accumulated during the past 4,000 years are in general small and sparse (Cordova et al. 2005). Originally named the "Younger Fill" by Claudio Vita-Finzi (1969), this deposit encompasses deposits of various ages ranging from Nabataean to Early Islamic (Cordova et al. 2005). This alluvial fill occupies the bottom of the valleys, where it is overshadowed by the magnitude of alluvial fills of earlier age (see fig. 5.6). Although present in many wadis of Jordan, the causes of the deposition of this historic alluvial unit have not been studied.

The lack of sizable changes in the stream networks of the Jordanian wadis during the past 4,000 years can be seen in the construction and relative preservation of water management installations. In most cases, modifications of the stream network were created by damming streams and other works of land reclamation during the Nabataean, Roman, Byzantine, and Islamic periods. Dams and gristmills dating to this period are found in many areas of western Jordan (Walker 1999). However, these modifications were localized and had little impact on the streams.

Unlike the Early and Middle Holocene, the Late Holocene saw no major environmental crisis. However, minor and more localized crises occurred in several parts of the Levant, including Transjordan, multiple times. It is important, for example, to mention the decline of settlement at the end of the Late Bronze Age and during the early centuries of the

Islamic period. In these two cases, however, the effects of climate are difficult to discern because of the influence of socioeconomic factors. These crises are mainly characterized by changes in the archaeological record but not in geoarchaeological and paleoecological records.

Final Consideration on Historic Landscape Transformation

Considering the four premises above, it is possible to believe that even the peoples who inhabited Transjordan during the Iron Age faced an already deteriorated environment. Thus, the land reclaimed by the Nabataeans may have looked very much like the landscape that modern Bedouins are settling today. Perhaps the peoples of the Neolithic were the only ones who were able to see Transjordan as a real paradise with abundant soil and diverse flora and fauna. This relationship between landscape, land degradation, and time is what I refer to as landscape inheritance. Each group during each of the cultural periods inherited problems (i.e., eroded soils, degraded vegetation) from previous periods. Because of the relatively dry climate and variable rainfall, the period of vegetation and soil resilience is lengthy. Consequently, I apply the concept of inherited landscape to a landscape that a generation receives from their predecessors, independently of the degree of resilience. In the case of the post-Neolithic landscapes in the Levant, inherited landscapes handed down to younger generations became more degraded as time progressed.

Whether in good or bad shape, the inherited landscape plays an important role in the possibilities of land use and resource exploitation. Therefore, this concept should be considered when evaluating the environmental adaptation strategies of an ancient society.

7

Interpreting Millennial Landscape Change

Woodlands, Steppes, and Deserts over the Millennia

Pollen records from lake deposits of the Rift Valley, along with numerous sets of archaeobotanical, archaeozoological, and pedological data, indicate that the limits of the floristic regions in Transjordan varied considerably during the past 20,000 years (see fig. 6.1). The wooded areas of the Western Highlands reached their maximum extension around 13,000 years ago, which is by far the largest expansion of wooded landscapes during the past twenty millennia. This process was hindered first by the relatively short cold-dry event of the Younger Dryas and later by the advent of farming and pastoralism during the Pre-Pottery Neolithic B period. Subsequently, wooded areas were progressively reduced as the rapid increase in population during the Chalcolithic and Early Bronze Age demanded more farming areas and timber. Thus, by 4,000 years ago, the areas covered with woods in Jordan were becoming smaller, perhaps similar to their present extent.

In addition to woodland reduction, the Holocene was also characterized by a change in vegetation composition. As a result of increasing grazing, low scrub (garrigue or *batha*) vegetation began to expand in areas of Mediterranean climate at the expense of other vegetation communities probably beginning in the Pre-Pottery Neolithic period. The reduction of wooded areas occurred along with the expansion of low scrub and the invasion of steppe plants that led to the formation of the nonforest Mediterranean vegetation Dawud al-Eisawi described (1985).

The remaining woodlands are today located in sanctuaries concentrated in two large areas in the north and south of the Transjordanian Mediterranean Belt (see fig. 3.1). The northern wooded area is located around the Ajlun Mountains and the escarpment facing the Jordan Valley and the Yarmouk Valley. The southern woods are clustered in the Tafila Highlands and the Dana Nature Reserve. Small relict communities of trees, especially pistachio, are scattered in several areas of the non-forested Mediterranean region, as is the case of the Ataruz area on the western margin of the Madaba Plateau.

The Irano-Turanian steppes also underwent profound changes during the Holocene. During the Early Holocene, a grass-forb steppe was replaced by the present scrub steppe dominated by Chenopodiaceae (mainly the genera *Anabasis* and *Noaea*), sages (*Artemisia*), and other antipastoral plants. Intense soil erosion occurred, especially in areas of loess. The invasion of desert plants and the overall pathetic aspect of the Irano-Turanian steppe vegetation make it sometimes difficult to distinguish from the Saharo-Arabian desert vegetation.

The Saharo-Arabian and Sudanian vegetation regions were also deeply transformed during the Holocene, despite the low population levels in them. After the Last Glacial Maximum, these desert regions underwent rapid wet-and-dry cycles as well as changes in seasonal precipitation regimes. Increased temperatures and humidity, possibly enhanced by summer monsoonal rains, favored the expansion of tropical Sudanian elements on at least two occasions, around 13,000 and 8,000 years ago (see fig. 6.1). These warm and moist events resulted from an increase in average temperatures and the expansion of the Indian Ocean Monsoon (Roberts and Wright 1993; Bar-Matthews, Ayalon, and Kaufman 2000; Issar 2003). However, this phenomenon still needs to be investigated in terrestrial records, given the controversial data suggesting that rain increase in the Early Holocene did not involve monsoon rains (Arz et al. 2003).

The dry events that occurred during the Terminal Pleistocene, particularly the Younger Dryas, contributed to reduce the amount of water that fed the dozens of lakes that existed in the Central Plateau during the Pleistocene (see fig. 5.9). Thus, the drying lakes became the qaʿat (dry playa basins) typical of the Jordanian deserts. This drying process led to the shrinking of Lake Lisan into the Holocene Dead Sea.

Myths and Realities about the Jordanian Landscapes

There is no doubt that the territory of Jordan and the entire Levantine region were profoundly transformed by humans, as originally pointed out by the geographer George Perkins Marsh in the late 1800s (Lowenthal 2000). Marsh's ideas influenced geographers during the early half of the twentieth century, furthering a series of explanations for the processes that led to the modern dry and unproductive landscapes of the Levant. Ellsworth Huntington (1911) blamed it mainly on deteriorating climate, while Walter Clay Lowdermilk (1944) and Adolf Reifenberg (1955) attributed it to the nomadization of the rural economy. The latter pointed to a general landscape deterioration that resulted from land mismanagement during the Early Islamic period. According to this scheme, local sedentary populations turned to nomadism, which in turn led to the increase of grazing that deteriorated the land. The ideas of these scholars stressed the negative ecological aspects of nomadism as the opposite to and adversary of sedentary life, which eventually evolved into the so-called Desert and Sown myth. The ideas surrounding this myth would prevail in the archaeological and geographical literature for several decades until landscape reconstruction studies based on empirical data began to yield results that contradicted most of the beliefs grounded in the Desert-Sown dichotomy.

The picture portrayed by the Desert-Sown dichotomy is obsolete in the concept of landscape degradation (see chapter 1). Although the land has been used and forced into deterioration, it has also undergone some form of resilience through natural processes (climatic amelioration) and cultural processes (land reclamation). The recovery of landscapes to a state close to their original (preagricultural) conditions has never been achieved. Nonetheless, landscape structure in protected areas (e.g., Dana and Shaumari reserves) has been restituted such that one day they may achieve the appearance of preagricultural landscapes, although reaching this point may take a very long time.

The low rates of resilience in the semiarid and arid areas of the Levant should be taken into consideration when evaluating the potential of landscapes that had previously maintained farming and pastoralism. Therefore (contra Gophna, Liphschitz, and Lev-Yadun 1986), paleolandscape modeling using settlement-based population levels for post-Neolithic periods should not consider a start point in a fully wooded landscape with abundant soil.

At the beginning of the twenty-first century, numerous studies on landscape reconstruction have deeply modified our conceptions about landscape degradation in the Levant. Such studies include a variety of sources, including fossil pollen data, alluvial chronologies, soil erosion data, archaeobotanical and paleobotanical data, and stable isotope data, among others. All of the evidence recovered through these studies is still minimal, especially for Jordan. Therefore, students of archaeology and geography who are interested in the Near Eastern landscapes should be encouraged to pursue research focusing on the recovery of paleoecological proxy data, geoarchaeological research, and interpretation of modern and past cultural ecological processes.

The rapid and extreme climatic changes following the Last Glacial Maximum (i.e., after ca. 20 ka BP) impacted the landscapes of Transjordan in many ways. Large amounts of red soils were removed from the plateaus before agriculture even began (Cordova 2000). After these changes, the landscape was always prompt to exhibit resilience. However, after 10 ka BP, the resilience process was interrupted by a form of environmental degradation more acute than the one produced by climate change. The human-induced changes in the Holocene occurred faster and were more intense, leaving very short times for landscape resilience. This observation led me to put forward the concept of inherited landscape, discussed at the end of chapter 6.

Threshold Crossing and Environmental Crises

The evolution of inherited landscape can be explained using the concept of threshold, in a sense similar to the one in fluvial geomorphology. The concept of threshold implies that if a certain limit is crossed, the new conditions achieved by a system would be irreversible. The notion of crossing a threshold in the process of landscape transformation implies that changes inflicted upon the landscape may exceed any possibilities of resilience.

A simple case of threshold crossing is exemplified by the effects of intense soil erosion. If soils are removed by erosion after a moderate removal of vegetation, the original plant communities can recolonize the damaged area only if the conditions of the soils damaged by erosion are still similar to those of the pre-erosional period. But if the processes of soil removal are pushed so far as to expose the underlying bedrock, then the original type of vegetation will not be able to recolonize the affected area. Thus, new pioneer and invasive flora better adapted to the new soil conditions will take over the eroded area. In this case a threshold, or a point of no return, has been crossed.

Other examples of threshold in landscape transformation include the total removal of native fauna. Throughout prehistoric periods, gazelles were hunted for a very long time but not to the point of extinction. In more recent times, gazelle hunting went far beyond the threshold, such that gazelles became extinct in Transjordan. However, the conditions beyond the threshold have been reversed because gazelles have now been reintroduced in some of the nature reserves in the country. Nonetheless, their ecological context is not the same as before, because this time they are kept within the boundaries of the nature reserves, where they are protected from predators. That their ecological context is different from the original implies that the conditions favorable to gazelle populations have not been fully reversed from beyond the threshold.

During the Holocene, the transformation of the Jordanian landscapes crossed several thresholds, resulting in the formation of the modern landscapes that bear little if any resemblance to the preagricultural landscapes. The crossing of these thresholds occurred during environmental crises, which here are divided into major and minor. The major environmental crises are large-scale disruptions in natural and environmental patterns of the landscape. During a major environmental crisis, natural systems such as soil, vegetation, and streams are pushed beyond their resilience limits. Minor environmental crises, by contrast, consist of small-scale disruptions of systems, sometimes affecting only one or two components of natural or cultural systems. Usually, major environmental crises have clear expression in archaeological, geoarchaeological, and paleoecological records. Minor environmental crises are not clearly defined or are defined in only one of the three types of records.

Two major environmental crises occurred during the Holocene: the one that occurred at the end of the Pre-Pottery Neolithic B (PPNB) and the one at the end of the third millennium BC. Both had a widespread regional impact, and both are clearly defined in archaeological, geoarchaeological, and paleoecological records, as demonstrated in chapter 6.

The crisis at the end of the PPNB is perhaps the most devastating and prolonged of the two major crises, and the one that pushed the limits of landscape resilience beyond their thresholds. As explained in chapter 6, the process of land use intensification, begun in the Early PPNB, reached its peak in the Middle PPNB. The Late PPNB and PPNC experienced shrinking of settlements, and as the paleobiological records indicate, faunal and floral diversity were dramatically reduced.

The impact of the PPNB-PPNC environmental crisis can be compared with the crisis that occurred in the Americas at the end of the sixteenth century and beginning of the seventeenth century, a time that was characterized by depopulation, soil erosion, flooding, and economic crises (Butzer 1992; Denevan 1992). The introduction of livestock grazing, an activity never before experienced by the local landscapes, created a series of transformations in vegetation, soils, and human populations.

As in the sixteenth-century Americas, epidemics were a major problem during the PPNB in the Levant. For example, bone disease associated with tuberculosis has been reported from pathological studies on bone remains of the Late PPNB (Rollefson 1998). That tuberculosis appears in the Neolithic is not surprising, since goats, sheep, and cattle are vectors for the disease (Rollefson 1998, 120). Overall, the initial expansion of the nonforest Mediterranean vegetation and the scrub steppe began after this crisis.

The major environmental crisis that occurred at end of the third millennium BC was caused by a complex combination of climatic and socioeconomic factors, as discussed in chapter 6. The response of population and landscape to the regional crisis varied throughout the Near East. For example, some settlements are known to have remained occupied during the time of the crisis. In some cases, however, such settlements eventually collapsed. Based on the model developed by Wilkinson (1994, 1997) in the Khabur Plains of Upper Mesopotamia, it is possible also to factor in aspects of population growth and carrying capacity to explain urban decline in a region. In the case of western Transjordan, the rapid growth of population and towns led to pressures on the environment that for some time were not evident, but as the economic crisis became more acute and the climate became drier, the decaying social systems advanced to their eventual collapse.

It has been hypothesized that most Early Bronze Age towns in Transjordan and Palestine depended on floodplain irrigation, which is assumed on the grounds of their location near streams (Mabry 1992). The shift to arid conditions led to irreversible changes in the streams that resulted in widespread stream incision. As a result, the water table dropped and flash-flood regimes eroded the floodplain. The geoarchaeological study of soils and sediments around Khirbet Iskander seems to support this scenario, which has been proposed for areas elsewhere in the Levant (Mabry 1992; Rosen 1997a) and Upper Mesopotamia (Rosen 1997b; Wilkinson 1997, 1998, 1999).

Examples of minor environmental crises that occurred in Transjordan during the Holocene include several, but the best known are the climatic crisis that occurred in the early third millennium BC (Weiss 2000), the complex and less understood crisis at the beginning of the first millennium BC, the crisis that occurred under social changes and dry conditions in the sixteenth century AD, and the crisis driven mainly by socioeconomic decline at the end of the Ottoman period (late nineteenth and early twentieth century).

We may also include among the minor crises the one that slowly developed during the Early Islamic period (ca. AD 700–1200), which was triggered not by deteriorating climate but by cultural changes that transformed aspects of social life, including settlement patterns and land use. However, the processes of landscape change were not as acute as Lowdermilk (1944) and Reifenberg (1955) proposed. Alexandre Kedar (1985) presents evidence that several social, economic, and ecological improvements were actually achieved in certain localities of the Levant during this period. Pollen records show an increase in forested areas (Schwab et al. 2004), which is not necessarily reflecting high precipitation levels but may also indicate a reduction of farmed areas.

The environmental crisis at the end of the Ottoman period had various natural and political components. Unlike the previous crises, it is best known through written information and climatic records (Abujaber 1989). The late Ottoman period (ca. 1800–1918) was characterized by widespread economic decline, evident in high taxation, droughts, Bedouin raids, and local tribal uprisings (Mousa 1982; Salibi 1998; Abujaber 1989; Walker 1999). The pathetic look of the Transjordanian landscape of that time is palpable in most of the descriptions by travelers, in particular Huntington (1911), who at that time attributed the deplorable looks of the landscape to climatic change, conflicts with the conclusion by Lowdermilk (1944) that attributed land deterioration to the increase in nomadism. Raouf Abujaber (1989) provides the best historical accounts of the Late Ottoman crisis in Transjordan by carefully analyzing its climatic and social causes. Accordingly, the crisis seems not to have been as bad as it was portrayed by Western travelers, who saw it through a foreign lens. The Ottoman administration, for instance, did as much as possible to settle farmers from Palestine and Syria in areas of fertile red soils on the plateaus of Madaba and Karak. Nevertheless, a high-resolution pollen diagram (Schwab et al. 2004) does not show substantial change in vegetation during this period as compared with the immediately previous periods. Therefore, the effects of this crisis still need to be analyzed with more detail.

Millennial Landscape Change Phases

The process of landscape change during the Terminal Pleistocene encompasses mainly climatic changes that followed the Last Glacial Maximum. They are represented mainly by peaks of warm-moist and dry-cold phases, culminating with the Younger Dryas around 11 ka BP. Afterwards, changes were largely the result of the interplay between climatic and human factors, which interacted to drive the antagonistic forces that shape the process of land degradation (e.g., erosion and soil formation; vegetation destruction and regrowth).

The main phases of landscape change during the Holocene include the transformation that occurred during the Neolithic through plant and animal domestication, the rise of urbanism during the Early Bronze Age, the unprecedented expansion of agriculture during the Roman-Byzantine period, and the changes that occurred in the industrial era, namely, after the end of the Ottoman period.

Environmental crises, prompted by climatic and human-induced deterioration processes, forced the landscape across various thresholds. These thresholds created nonreturn changes despite resilience processes (natural and human driven). Landscape transformation in twenty-first-century Jordan, produced mainly through urbanization and commercial agriculture, is pushing the landscape beyond thresholds never crossed before. However, while some areas are deeply transformed, others have experienced resilience, which in turn creates reversals in the process of land degradation. Major reversals in the process of land deterioration are now being observed in the nature preserves established in the late twentieth century, for example. However, because of their short existence, it is too early to use them to interpret landscape resilience. The truth is that despite all conservation and protection practices, the territories within the reserves will never reach any close resemblance to prefarming and prepastoral landscapes. Nonetheless, the slowly resilient territories of the nature preserves, particularly those with reintroduced herbivores, could provide data in the form of vegetation structure, pollen, and phytoliths that could be used for the interpretation of paleoecological data for landscape reconstruction.

Appendix
Latin, English, and Arabic Names of Plants

Trees and Shrubs

Acacia (Fam. Leguminosae)
Acacia nilotica [(L.) Willd ex Delile], Nile acacia (E), kharūb miṣrī (A)
Acacia tortilis [(Forskal) Hayne], acacia (E), talḥ, sayāl (A)
Acacia tortilis subsp. *raddiana* [(Forsk.) Hayne (Savi) Brenan], acacia (E), talḥ (A)
Amygdalus (Fam. Rosaceae)
Amygdalus arabica [Oliv.], Arabian almond (E)
Amygdalus communis [L.], almond (E), lawz (A)
Amygdalus korschinskyi [(Hand-Mazzetti) Bornm.]
Arbutus (Fam. Ericaceae)
Arbutus andrachne [L.], madrone, strawberry tree (E), quṭlub (A)
Ceratonia (Fam. Fabaceae)
Ceratonia siliqua [L.], carob tree (E), kharawb (A)
Cercis (Fam. Fabaceae)
Cercis siliquastrum [L.], Judas tree (E), khirzīj (A)
Cistus (Fam. Cistaceae)
Cistus villosus [L.], pink rockrose (E), lubād shaʿrī (A)
Crataegus (Fam. Rosaceae)
Crataegus aronia [Bosc. Ex DC.], hawthorn (E), zaʿrūr (A)
Crataegus azarolus [Auct. et Boiss], hawthorn (E), zaʿrūr (A)
Cupressus (Fam. Cupressaceae)
Cupressus sempervirens [L.], Italian cypress (E), sarū (A)
Ficus (Fam. Moraceae)
Ficus carica [L.], common fig (E), tīn (A)
Ficus pseudosycamorus [Decne.], wild fig (E), tīn barrī (A)
Juniperus (Fam. Cupressaceae)
Juniperus phoenica [L.], Phoenician juniper (E), ʿarʿar (A)
Loranthus (Fam. Loranthaceae)
Loranthus acaciae [Zuc.], acacia strap flower (E), sittir, ʿannāb (A)
Moringa (Fam. Moringaceae)
Moringa peregrina [(Forskal) Fiori], hesaban (E,A), yusur (A)

Nerium (Fam. Apocynaceae)
Nerium oleander [L.], oleander (E), diflah (A)
Olea (Fam. Oleaceae)
Olea europaea [L.], olive tree (E), zaytūn (A). The domesticated variety is often referred to as *Olea europaea* var. *europaea*. The escaped or feral variety is *O. europaea* var. *oleaster*.
Phoenix (Fam. Arecaceae)
Phoenix dactylifera [L.], date palm (A), nukhayl, nakhīl (A)
Pinus (Fam. Pinaceae)
Pinus halepensis [Mill.] (also *P. hierosolimitana* [Duh.]), Aleppo pine, Jerusalem pine (E), ṣanawbar ḥalab (A)
Pistacia (Fam. Anacardiaceae)
Pistacia atlantica [L.], Atlas pistachio tree, terebinth (E), buṭm aṭlasī (A)
Pistacia khinjuk [Stocks], green pistachio (E), buṭm akhdar (A)
Pistacia palaestina [(Boiss) Post.], Palestinian pistachio (E), buṭm filasṭīnī (A)
Quercus (Fam. Fagaceae)
Quercus calliprinos [Webb.] (also *Q. coccifera* [L.]), Kermes oak (E), sindiyan, balūṭ mustadīn al-khudra (A)
Quercus ithaburensis [Decne.] (also *Q. aegylops* [L.]), tabor oak (E), mallūl, balūṭ mutasāqaṭ al-awrāq (A)
Retama (Fam. Leguminosae)
Retama raetam [(Forskal) Webb et Berth.], white broom, juniper bush, retam (E), ratam (A)
Rhamnus (Fam. Rhamnaceae)
Rhamnus palaestinus [Boiss.] (also *R. punctata* var. *palaestinus* [Post. A.]), Palestine buckthorn (E), suwwayd (A)
Salvadora (Fam. Salvadoraceae)
Salvadora persica [L.], salvadora (A), arāk, siwāk (A)
Styrax (Fam. Styracaceae)
Styrax officinalis [L.], styrax tree (E), ʿabhar, ḥawz (A)
Tamarix (Fam. Tamaricaceae)
Tamarix amplexicaulis [(Ehrenb.)], tamarisk (E), ʾithal, ṭarfah (A)
Tamarix nilotica [(Ehrenb.) Bunge], tamarisk (E), ʾithal, ṭarfah (A)
Tamarix passerinoides [Delile ex Desv.], tamarisk (E), ṭarfah (A)
Tamarix tetragyna [Ehrenb.], tamarisk (E), ṭarfah (A)
Ziziphus (Fam. Rhamnaceae)
Ziziphus lotus [(L.) Lam.]
Ziziphus spina-christi [(L) Willd.], Christ's thorn (E), nabq, ṣidr (A)
Zygophyllum (Fam. Zygophyllaceae)
Zygophyllum dumosum [Boiss.], bushy bean caper (E), ʿadhab (A)

Herbs, Scrub Plants, and Short Shrubs

Alhagi (Fam. Fabaceae)

Alhagi maurorum [Medik.], camelthorn (E), ʿāqūl (A)

Anabasis (Fam. Chenopodiaceae)

Anabasis articulata [(Forskal) Moq.], jointed anabasis (E), ʿajram (A)

Anabasis syriaca [Iljin], Syrian anabasis (E), ʿaḍū (A)

Anthemis (Fam. Asteraceae)

Anthemis bornmulleri [Stoj. & Acht.], chamomile (E), ʾaqḥwān (A)

Anthemis maris mortui [Eig], Dead Sea chamomile (E), ʾaqḥwuān (A)

Anthemis melampodina [Delile] subsp. *deserti* [(Boiss) Eig], desert chamomile (E), ʾaqḥwuān (A)

Artemisia (Fam. Asteraceae)

Artemisia herba-alba [Asso] (known also as *A. inculta* [Delile]), herba-alba wormwood, sagebrush (E), shīḥ (A)

Artemisia judaica [L.], Judean wormwood (E), baʿīthrān (A)

Artemisia monosperma [Delile], sand wormwood (E), ʿādhir

Asphodelus (Fam. Liliaceae)

Asphodelus aestivus [Brot.], asphodel (E), ghuṣlān (A)

Astragalus (Fam. Fabaceae)

Astragalus bethlemiticus [Boiss.], Bethlehem vetch (E), jadād (A)

Astragalus spinosus [(Forskal) Muschler], milk vetch (E), kadād, qatād, jadād (A)

Atriplex (Fam. Chenopodiaceae)

Atriplex halimus [L.], tall orache, Spanish sea purslane (E), qaṭaf (A)

Atriplex holocarpa [F. Muell.] saltbush (E)

Ballota (Fam. Labiatae)

Ballota undulata [(Sieber ex Fresen.) Bentham], common black horehound (E), rasā, ʾasaghān (A)

Bellevalia (Fam. Hyacinthaceae)

Bellevalia spp. [Lapeyr.], bellevalia (E), baṣīl (A)

Calligonum (Fam. Polygonaceae)

Calligonum comosum [L'Her.], calligonum (E), ʾarṭaʾ (A)

Calligonum tetrapterum [Jaub et Spach], calligonum (E), ʾarṭaʾ

Calycotome (Fam. Fabaceae)

Calycotome villosa [(Poiret) Link], thorny broom (E), qandīl (A)

Capparis (Fam. Capparaceae)

Capparis cartilaginea [Decne.], caper (E), laṣaf (A)

Capparis leucophylla [DC.], caper (E), shafallḥ (A)

Capparis spinosa [L.], Egyptian caper (E), kabbār (A)

Carlina (Fam. Asteraceae)

Carlina hispanica [Lam.], corymbed carline thistle (E), sāq al-ʿarūs, zand al-ʿabd (A)

Carthamus (Fam. Asteraceae)
Carthamus tenuis [(Boiss. & Blanche) Bornm.], slender safflower (E), qūs (A)
Centaurea (Fam. Asteraceae)
Centaurea cyanoides [Berggren & Wahnleb], Syrian cornflower (E), shabbah (A)
Centaurea dumulosa [Boiss.], shrubby centaury (E), murrār shājīrī, katīlah (A)
Centaurea iberica [Trev. Ex Sprengel], Iberian centaury (E), murrār (A)
Colchicum (Fam. Liliaceae)
Colchicum shemperi [Janara], autumn crocus (E), 'islān (A)
Cressa (Fam. Convolvulaceae)
Cressa cretica [L.], alkali weed (E), dhalulḥ (A)
Cynara (Fam. Asteraceae)
Cynara syriaca [Boiss.], Syrian artichoke (E), kharashūf barrī, arḍī shawkī (A)
Ephedra (Fam. Ephedraceae)
Ephedra transitoria [Riedl.], ephedra (E), kuddab (A)
Euphorbia (Fam. Euphorbiaceae)
Euphorbia hyerosolimitana [Boiss.], spurge (E), lubbīn, ḥalabūb, ḥalbabūn (A)
Evax (Fam. Asteraceae)
Evax contracta [Boiss.], cudweed (E), quṭīnah (A)
Fagonia (Fam. Zygophyllaceae)
Fagonia mollis [Del.], soft fagonia (E), ḥalūah (A)
Gomphocarpus (Fam. Asclepiadaceae)
Gomphocarpus sinaicus [Boiss.], gomphocarpus (E), ghaliqat ad-dubb (A)
Gymnarrhena (Fam. Asteraceae)
Gymnarrhena micrantha [Desf.], gymnarrhena (E), khuf el-kalb (A)
Halogeton (Fam. Chenopodiaceae)
Halogeton alopecuroides [(Delile) Moq.], halogeton (E)
Haloxylon (Fam. Chenopodiaceae)
Haloxylon articulatum [(Cav.) Bunge], salt tree (E)
Haloxylon persicum [Bunge], haloxylon (E), ghaḍā (A)
Hammada (Fam. Chenopodiaceae)
Hammada eigii [Iljin], hammada (E), ḥamaḍ (A)
Hammada salicornica [(Moq.) Iljin], hammada (E), ḥamaḍ, ramth (A)
Hammada scoparia [(Pomel) Iljin], hammada (E)
Iris (Fam. Iridaceae)
Iris nigricans [Dinsm.], black iris (E), sawsan 'aswad (A). Jordan's national
    flower.
Marrubium (Fam. Labiatae)
Marrubium libanoticum [Boiss.], horehound (E)
Marrubium vulgare [L.] var. *lunatum* [Benth.], horehound (E)
Medicago (Fam. Fabaceae)
Medicago laciniata [(L.) Miller], cutleaf medick (E)
Nitraria (Fam. Zygophyllaceae)

Nitraria retusa [(Forsk.) Ascherson], nitraria (E), gharqal (A)

Noaea (Fam. Chenopodiaceae)

Noaea mucronata [(Forskal) Ascheson & Schweinf], thorny saltwort (E), ṣirr (A)

Notobasis (Fam. Asteraceae)

Notobasis syriaca [(L.) Cass.], Syrian thistle (E), kharafīsh al-kabīr (A)

Onopordon (Fam. Asteraceae)

Onopordon alexandrinum [Boiss.], cotton thistle (E), ʿattūr (A)

Ornithogalum (Fam. Liliaceae)

Ornithogalum narbonense [L.], star of Bethlehem (E), laban aṭ-ṭayr an-narbunī, nijmat bayt laḥm (A)

Panicum (Fam. Poaceae)

Panicum turgidum [Forskal], tussock grass, desert grass (E), taman, tuman, thaman (A)

Peganum (Fam. Zygophyllaceae)

Peganum harmala [L.], peganum (E), ḥarmal (A)

Pergularia (Fam. Asclepiadaceae)

Pergularia tomentosa [L.], pergularia (E), ghaliqah (A)

Picnomon (Fam. Asteraceae)

Picnomon acarna [(L.) Cass.], welted thistle, yellow cnicus (E), shawk al-fār (A)

Plantago (Fam. Plantaginaceae)

Plantago cylindrica [Forskal], plantain (E)

Plantago lanceolata [L.], narrowleaf plantain (E)

Plantago ovata [Forskal], psyllium seed (E)

Poa (Fam. Poaceae)

Poa bulbosa [L.], bulbous meadow grass (E), nazaʿa (A)

Poa sinaica [Sleud], Sinai meadow grass (E), nazaʿa ansayʿah (A)

Polypogon (Fam. Poaceae)

Polypogon monspeliensis [(L.) Desf.], annual rabbitsfoot grass (E)

Rumex (Fam. Polygonaceae)

Rumex acetosella [L.], common sheep sorrel (E)

Salsola (Fam. Chenopodiaceae)

Salsola baryosma [(Schultes) Dandy], fetid saltwort (E), wuqqīd, kharīd (A)

Salsola vermiculata [L.], narrow-leaved saltwort (E), ḥamaḍ (A)

Salvia (Fam. Labiatae)

Salvia dominica [L.], Dominica sage (E), marrū, khuykah (A)

Sarcopoterium (Fam. Rosaceae)

Sarcopoterium spinosum [(L.) Spach], also known as *Poterium spinosum* [L.], prickly shrubby burnet (E), natish blān (A)

Scolymus (Fam. Asteraceae)

Scolymus maculatus [L.], spotted golden thistle (E), sunnārīyah (A)

Scorzonera (Fam. Asteraceae)

Scorzonera judaica [Eig], yellow viper's grass (E), qaʿfūr, arīhā (A)

Scorzonera papposa [DC.], pink viper's grass (E), nibbaḥ (A)
Seidlitzia (Fam. Chenopodiaceae)
Seidlitzia rosmarinus [Boiss.], seidlitzia (E), shanānah (A)
Silene (Fam. Caryophyllaceae)
Silene villosa [Forskal], villous catchfly (E), nawuār, ʿaṭānī (A)
Silybum (Fam. Asteraceae)
Silybum marianum [(L.) Gaertner], holy thistle (E), kharafīsh jimāl (A)
Spergularia (Fam. Caryophyllaceae)
Spergularia diandra [(Guss.) Heldr. & Start.], diandrous spurrey (E), ʾumm tharīb, ʾabū ghullah (A)
Spergularia marina [(L.) Griseb.], sea spurrey (E), ʾumm thurayb (A)
Spergularia salina [J. & K. Presl], salt sandspurrey (E)
Stipa (Fam. Poaceae)
Stipa capensis [Thunb.], twisted feather (E), safsuf (A)
Suaeda (Fam. Chenopodiaceae)
Suaeda asphaltica [(Boiss.) Boiss], asphaltic sea-blight (E)
Suaeda vera [J. F. Gmelin], shrubby sea-blight, seepweed (E), suwwyīd, ḥimām (A)
Traganum (Fam. Chenopodiaceae)
Traganum nudatum [Delile], traganum (E), zamrān, ḥamḍ, swayd (A)
Urginea (Fam. Liliaceae)
Urginea maritima [(L.) Baker], squill (E), ʿislān, buslān, bisūl (A)
Urtica (Fam. Urticaceae)
Urtica spp. [L.], nettle (E), qurrīṣ (A)
Varthemia (Fam. Asteraceae)
Varthemia iphionoides [Boiss. & Blanche], common varthemia (E), kuttīla, ṣaf-fīrrah (A)

Aquatics and Semiaquatics

Arundo (Fam. Poaceae)
Arundo donax [L.], giant reed (E), ghāb rūmī, ghāb baladī, qaṣab (A)
Imperata (Fam. Poaceae)
Imperata cylindrica [(L.) Beauv.], blady grass, sword grass (E)
Inula (Fam. Asteraceae)
Inula chrithmoides [L.], golden samphire (E), ḥaṭab zītī (A)
Juncus (Fam. Juncaceae)
Juncus acutus [L.], Leopold rush (E)
Juncus bufonius [L.], toad rush (E), samār (A)
Juncus maritimus [L.], sea rush (E), samār (A)
Lemna (Fam. Lemnaceae)
Lemna gibba [L.], duckweed (E)

Phragmites (Fam. Poaceae)
Phragmites australis [(Cav.) Trin. ex Stendel], phragmites (E), ghāb (A)
Ruppia (Fam. Ruppiaceae)
Ruppia cirrhosa [Petagna], widgeon weed, ditchgrass (E)
Ruppia maritima [L.], widgeon grass (E)
Scirpus (Fam. Cyperaceae)
Scirpus maritimus [L.], seaside bulrush, sea club-rush (E)
Sonchus (Fam. Asteraceae)
Sonchus maritimus [L.], sea-sow thistle (E), juʿdīd, ʿilk (A)
Sonchus oleraceus [L.], common sow thistle (E), ʿilk khayl (A)
Typha (Fam. Typhaceae)
Typha angustata [Boryet Chaub.], cattail (E), ḥalfā (A)
Typha domingensis [Pers], cattail (E), ḥalfā (A)
Zanichellia (Fam. Zanichelliaceae)
Zanichellia palustris [L.], horned pondweed (E)

Glossary

Acheulian (or Acheulean). Lithic tool-making tradition dated from around 1,500,000 to 200,000 years ago. Although the name comes from St. Acheul in France, Acheulian lithics are found in many localities in Europe, Africa, and Asia. This technological tradition includes the making of bifaces (handaxes), picks, and cleavers (square-ended tools). These tools are generally accepted to have been used by *Homo erectus*.

aggradation. The process of sediment accumulation that causes a surface to rise. Stream aggradation causes the channel to fill up with sediment and the level of a floodplain to rise. The term is also applied to the process of silting-up in a lake or a reservoir.

algal stromatolites. Colonial structures, usually calcareous, that are formed by algae in subtidal and shallow water environments.

antipastoral. Adjective referring to plants avoided by livestock because they are unpalatable or poisonous to livestock. The garrigue vegetation is an example of an antipastoral plant community.

argillic horizon. Mineral soil horizon containing clay. Usually clay accumulations result from illuviation (see also illuvial horizon). In the rendering of a soil profile, argillic horizons are indicated with the designation "Bt."

aridic moisture regime. Moisture regime typical of soils in arid lands. The soil experiences not less than 90 consecutive days when the temperature is above 8°C (46.4°F) and is dry more than one-half of the time when the temperature is above 5°C (41°F).

barchan dune. Crescent-shaped sand dune, where the horns point in the direction of the wind. Slope is gentle on the windward side of the dune and steep on the leeward side.

barchanoid dune (also barchanoid ridges). Ridge dune formed by several interconnected barchan dunes.

batha. Equivalent of garrigue in Israel.

breccia. Rock formed by angular fragments of rock. Breccias can be of volcanic (pyroclastic) or sedimentary origin.

calcic horizon. Mineral soil horizon of secondary carbonate enrichment that is at least 15 centimeters thick and has at least 5 percent or more $CaCO_3$ (or

5 percent more by volume than the underlying horizon). In the rendering of a soil profile, calcic horizons are indicated with the designation "Bk."

caliche. Mexican-Spanish term of the American Southwest used to designate a calcrete. Its equivalent in Jordan is *nārī*.

cambic horizon. Mineral soil horizon located at least 25 centimeters deep, with stronger chroma or redder hues than in the underlying horizon, with structure different from the underlying sediment or rock, and with little evidence of illuviation. In the rendering of a soil profile, cambic horizons are indicated with the designation "Bw."

chert. Siliceous rock composed of opaline and chalcedony silica. Chert forms in bands along bedding planes in limestone formations. Chert is black or dull in color and breaks off in splints and planes. The terms *flint* and *chert* are often used indistinctively because of their similarities in properties and use.

climbing dune. Ramp of eolian sand accumulated against a cliff. Climbing dunes are located on the windward side of the obstacle.

conglomerate. A consolidated sedimentary deposit or rock composed of rounded and subgrounded rock fragments such as pebbles and cobbles in a matrix of finer material.

continentality. Set of climatic conditions characteristic of inland regions with low exposure to the influence of air from seas and oceans. The main characteristics of a continental climate are low air humidity, low precipitation, and strong differences between mean July and January temperatures.

echo dune. Ramp of eolian sand accumulated on the lee side of an obstacle.

ecotone. The transitional zone between two ecosystems. For example, the transition between forests and steppes is marked by an ecotone that has elements of both.

Emberger quotient (Q). A pluviothermic quotient created by Louis Emberger in 1939, designed to subdivide the Mediterranean climates into four subtypes: arid, semiarid, subhumid, and humid. The determination of each subtype is obtained by using a two-dimensional diagram where the average minimum temperature of the coldest month is the X axis and the pluviothermic quotient is the Y axis. The climate subtype is determined according to the zone in which the X and Y values converge in the diagram. The Emberger quotient (Q) is estimated with the following formula:

$Q = P/0.5 \ (M + m) \ (M - m)$, where
P = annual rainfall in millimeters
M = average of the maximums of the hottest month (°C)
m = average of the minimums of the coldest month (°C)

endorheic basin. Drainage basin with no outlet to the sea. Usually water is collected in a lake or playa.

exorheic basin. Drainage basin with streams that eventually reach the sea.

facies (sedimentary facies). A group of features that reflect the specific environmental conditions in which a sedimentary deposit or rock formed. For example, fluvial facies could be channel, overbank, backswamp, etc. The features may be lithological, sedimentological, or faunal.

fan-delta. An alluvial fan where the distal part ends below the waters of a lake or sea.

flint. Fine-grained siliceous rock similar to chert. In general, flint has a waxy appearance and presents lighter colors. Flint breaks in conchoideal (curvilinear) shapes, as opposed to the flatter shapes of chert. However, the differences with chert are so small that sometimes the terms *chert* and *flint* are used indistinctively.

formation (geological term). A distinctive body of rock or sediment that is well differentiated from underlying, overlying, and adjacent formations. Formations vary in size and thickness, but they can be mapped and traced for long distances. A formation is further subdivided into members.

garrigue. Low scrub plant community typical of overgrazed areas in the Mediterranean region. The composition of plants varies from low thorny shrubs to aromatic plants. Garrigue plants are usually unpalatable to livestock.

halophyte. Plant adapted to grow on saline soil.

hiatus (depositional hiatus). A gap in a sequence of strata, where the missing strata were never deposited or were destroyed by erosion before younger strata were deposited. In a stratigraphic sequence, a depositional hiatus is expressed in the form of an unconformity.

illuvial horizon. A mineral soil horizon in which fine material carried from an overlying layer has been precipitated from solution or deposited from suspension. The material deposited in this horizon includes clays and organics.

illuviation. The process of deposition of soil material from one horizon to the underlying horizon.

incision (stream incision). The process of sediment down-cutting by the erosive power of water in a stream channel. This process is also known as channel entrenchment. When stream incision occurs, a valley bottom becomes an alluvial terrace.

inselberg. A German word meaning "island or isolated mountain" that has been adopted to designate a prominent steep-sided hill of rock rising abruptly on a desert plain.

interglacial. A warm period between two glacials.

interstadial. A warm interval of time within a cold (glacial) period.

Levallois. A flaking technology that involves a substantial working/shaping of the core to remove a flake, blade, or point of predetermined size and shape. For many archaeologists it demonstrates a level of forethought and planning not seen before in the archaeological record. It is traditionally associated with Neandertals and the Middle Paleolithic, but an increasing body of evidence indicates that it dates back to the Acheulian (Lower Paleolithic).

lunette. An accumulation of eolian sediment, often in the form of clay pellets the size of sand grains, formed around the lee margins of a playa lake.

maquis. Shrubland characterized by abundance of evergreen woody species. It is a typical plant community of the Mediterranean regions. The majority of the species are sclerophyl ("hard leaves"), such as certain species of oak, wild olive, and madrone, among others.

marine isotope stages (MIS). Often known also as OIS (ocean isotope stages), the MIS are time periods in the Quaternary defined by fluctuations in the curve temperature obtained from oxygen isotopes measured from sea-bottom sediments. Because the patterns of the temperature curve replicate in cores from various ocean regions of the world, the MIS are used as a global standard for paleotemperature fluctuations that mark long paleoclimatic periods (e.g., glaciations, interglacials, and interstadials).

member (geology). A subdivision within a formation. A member has lithologic properties different from other parts of the formation.

Mousterian. A Middle Paleolithic stone tool assemblage (ca. 250,000–40,000 years ago) found in the Levant and Europe composed primarily of side scrapers, notches, and denticulates. More rarely it also includes "Upper Paleolithic" tool types such as end scrapers, burins, borers, and backed knives. There are six variants of the Mousterian identified by the percentage/ratio of these and other stone tool types as well as its Levallois index. The Mousterian is primarily associated with Neandertals, but in the Levant it is also associated with anatomically modern humans while Neandertals in the Upper Paleolithic are producing Upper Paleolithic technologies.

nārī. Arabic word for caliche or calcrete.

nebkha (or nebka). Arabic term referring to a small accumulation of eolian sediment around an obstacle (rock or shrub). In North America, these features are known as "coppice dunes."

neotectonics. Term referring to tectonic movements that occur at present or occurred in the recent past. Neotectonic research involves measuring the rate of slippage of landmarks along faults often used for studying earthquakes.

normal fault. Fault in which displacement of a block occurs downwards along the hanging wall. In this case, the movement of the down-thrown block occurs along the vertical plane of a fracture.

optically stimulated luminescence (OSL). A radiometric dating technique based on the same principle as thermoluminescence. However, OSL is particularly used for the dating of sediments where exposure to light is limited. This technique measures the luminescence emitted from the most light-sensitive electron traps in particular minerals, especially quartz and feldspars, following exposure to light. Unlike thermoluminescence, in OSL, particles do not need exposure to heat to reset the radiation clock.

oxygen isotope ratio (OIR). The ratio between two stable isotopes of oxygen (^{18}O and ^{16}O). In water the OIR is often used as an indicator of water temperature and consequently of global and regional temperatures. This ratio is often expressed as $\delta^{18}O$.

pedogenesis. The natural process of soil formation.

pedogenic. Adjective referring to pedogenesis.

playa. Flat area occupying the center of a closed basin in desert areas. Temporary lakes occupy a playa, although most of the time it is characterized by crusts of silt and soluble salts. The interior of a playa is composed of layers of lacustrine and eolian sediments.

qaʿ. Playa, former lacustrine depression in the desert.

rubification. Reddening of soils through oxidation of iron. Although rubification can occur in several climates, it is very common in areas with cool wet winters and hot dry summers, i.e., Mediterranean climates.

ruderal. Plant species adapted to disturbed environments.

salt diapir. A salt dome formed by upward pressure created by underlying layers.

sapropel. Organic mud layer in sea-bottom sediments. Sapropels form as conditions in the sea water become stratified, so that vertical mixing is restricted. Through this process, the lack of oxygen allows no oxidation of the organic materials, and hence facilitates their preservation. In the Mediterranean, they are thought to be the result of a large influx of fresh water from large rivers. As fresh water enters the ocean it tends to float, consequently forming an upper water stratum, which restricts the vertical mixing of oxygen.

secondary carbonates. Carbonate accumulation in soils originated through processes of soil formation (pedogenic processes). Secondary carbonate morphology is usually accumulated in calcretes, in concretions or nodules, or along root casts and cracks.

speleothems. Crystalline formations that grow by continuous deposition of calcium carbonate, gypsum, and other soluble minerals. They are also known as stalactites and stalagmites, depending on their position with respect to the floor and ground of the cavern. Speleothems are layered, with each layer representing a phase of deposition and hence containing isotopic information related to the water that formed it. Usually, this water is meteoric water that seeps through cracks into the cavern, dissolving minerals on its way down.

strath terrace. A river terrace cut on bedrock, hence forming a bench on the valley slope.

strike-slip fault. Also known as tear-fault. A fault in which two sections move horizontally in opposite directions, that is to say, parallel to the plane of the fracture.

terra rossa. Reddish clay-loam sediment or soil rich in iron sesquioxides. There is no agreement regarding the origin of the terra rossa in the Mediterranean region. While some consider that terra rossa is residual material from the weathering of limestone, others consider that it is formed by dust particles.

thermoluminescence (TL). A radiometric dating method used on rocks, sediments, and pottery. Age determination is based on measurements of light emitted by certain mineral particles after being heated. Light emission occurs as that sample is exposed to continuous radiation after the heating ceased. Thus, thermoluminescence is the amount of radiation received per year (radiation dose rate).

unconformity (stratigraphic unconformity). A boundary separating two or more rocks or sedimentary units of markedly different age. An unconformity is the stratigraphic expression of a depositional hiatus.

wādī. Arabic word for "valley" and "stream." In geomorphic literature, the term has been adopted for designating a dry course of water, equivalent to the arroyos of the American Southwest.

xeric moisture regime. The moisture regime in which winters are moist and cool and summers are warm and dry. This is the typical moisture regime of the Mediterranean climate regions.

References

Abed, A. M. 1985. Paleoclimates of the Upper Pleistocene in the Jordan Rift. In *Studies in the history and archaeology of Jordan 2*, ed. A. Hadidi, 81–93. Amman: Department of Antiquities of Jordan.

———. 1998. Geology and mineral genesis of Taba inland sabkha, southern Dead Sea transform, SW Jordan. In *Quaternary deserts and climatic change*, ed. A. S. Alsharhan, K. W. Glennie, G. L. Whitle, and C. G. Kendall, 71–84. Rotterdam: A. A. Balkema.

Abed, A., P. Carbonel, J. Collina-Girard, M. Fontugne, N. Petit-Marie, J. C. Reyss, and S. Yasin. 2000. Un paléolac du dernier interglaciaire pléistocène dans l'extrême-sud hyperaride de la Jordanie. *Earth and Planetary Sciences* 330 (4): 259–64.

Abed, A. M., and F. F. Helmdach. 1981. Biostratigraphy and mineralogy of the Lisan Series (Pleistocene) in the Jordan Valley. *Berliner Geowissenschaft Abhandlungen* A 23:123–33.

Abed, A. M., and R. Yaghan. 2000. On the paleoclimate of Jordan during the last glacial maximum. *Palaeogeography, Palaeoclimatology, Palaeoecology* 160 (1–2): 23–33.

Abu-Irmaileh, B. 1988. *Poisonous plants in the environment of Jordan*. Amman: University of Jordan Press. (In Arabic)

Abu-Jaber, N. 1998. A new look at the chemical and hydrological evolution of the Dead Sea. *Geochimica et Cosmochimica Acta* 62 (9): 1471–79.

Abujaber, R. S. 1989. *Pioneers over Jordan: The frontier of settlement in Transjordan, 1850–1914*. London: I. B. Tauris.

Akazawa, T., K. Aoki, and O. Bar-Yosef, eds. 1998. *Neandertals and modern humans in western Asia*. New York: Plenum Press.

Albright, W. F. 1925. The Jordan Valley in the Bronze Age. *Annual of the American Schools of Oriental Research* 6:13–74.

Al-Eisawi, D. 1985. Vegetation in Jordan. In *Studies in the history and archaeology of Jordan 2*, ed. A. Hadidi, 45–48. Amman: Department of Antiquities of Jordan.

———. 1996. *Vegetation of Jordan*. Cairo: UNESCO Regional Office for Science and Technology for the Arab States.

Al-Eisawi, D., and B. Dajani. 1987. A study of airborne pollen grains in Amman, Jordan. *Grana* 26:231–38.

———. 1988. Airborne pollen of Jordan. *Grana* 27:219–27.

Al-Hunjul, N. G. 1995. *The geology of Madaba area: Map sheet (3153-II)*. Geological Bulletin 31. Amman: Geology Directorate, Geological Map Division, Natural Resources Authority.

Alley, R. B. 2000. The Younger Dryas Cold Interval as viewed from central Greenland. *Quaternary Science Reviews* 19 (1–5): 213–26.

Allison, R. J., N. J. Rosser, J. Warburtoin, J. R. Grove, D. L. Higgitt, A. J. Kirk. 2000. Geomorphology of the Eastern Badia Basalt Plateau, Jordan. *Geographical Journal* 166 (4): 352–70.

Al-Malabeh, A. 1994. Geochemistry of two volcanic cones from the intracontinental plateau basalt of Harra El-Jabban, NE Jordan. *Geochemical Journal* 28 (6): 517–40.

Aloni, E., A. Eshel, and Y. Waisel. 1997. The botanical conquest of the newly exposed shores of the Dead Sea. In *The Dead Sea: The lake and its setting*, ed. R. M. Niemi, Z. Ben Avraham, and J. R. Gat, 277–81. New York: Oxford University Press.

Amr, Z. S., G. Kalishaw, M. Yosef, B. J. Chilcot, and A. al-Budari. 1996. Carnivores of Dana Nature Reserve (Carnivora: Canidae, Hyenidae and Felidae), Jordan. *Zoology in the Middle East* 13:5–16.

Andrews, I. J. 1995. *The birds of the Hashemite Kingdom of Jordan*. Musselburgh, U.K.: I. J. Andrews.

Artzy, M., and D. Hillel. 1988. A defense of the theory of progressive soil salinization in ancient southern Mesopotamia. *Geoarchaeology* 3 (3): 235–38.

Arz, H. W., F. Lamy, J. Pätzold, P. J. Miller, and M. Prins. 2003. Mediterranean moisture source for an Early-Holocene humid period in the northern Red Sea. *Science* 300 (5616): 118–21.

Atkinson, K., and P. Beaumont, 1971. The forests of Jordan. *Economic Botany* 25 (3): 305–11.

Awawdeh, M. M. 1998. *Reconstructing the Quaternary sedimentary environment of Qaʿ Al-Habbabya, NE Jordan*. Master's thesis, Yarmouk University, Irbid.

Ayalon, A., M. Bar-Matthews, and A. Kaufman. 2002. Climatic conditions during marine oxygen isotope stage 6 in the eastern Mediterranean region from the isotopic composition of speleothems of Soreq Cave, Israel. *Geology* 30 (4): 303–6.

Baierle, H. U. 1993. *Vegetation und Flora im südwestlichen Jordanien*. Berlin: J. Cramer and Gebrüder Borntraeger.

Banat, K. M., and O. M. Obeidat. 1996. Notes on biogenic tufas associated with the Zerqa-Maʿin Hot Springs. *Carbonates and Evaporites* 11 (2): 213–18.

Banning, E. E. 2001. Settlement and economy in Wādi Ziqlāb during the Late Neolithic. *Studies in the history and archaeology of Jordan 7*, ed. Department of Antiquities of Jordan staff, 149–55. Amman: Department of Antiquities of Jordan.

Banning, E. E., R. Dods, J. Field, I. Kuijt, J. McCorriston, J. Siggers, H. Taani, and J. Triggs. 1992. Tabaqat al-Buma: 1990 excavations at a Kebaran and Late Neolithic site in Wadi Ziqlab. *Annual of the Department of Antiquities of Jordan* 36:43–69.

Bardtke, D. H. 1956. Die Waldgebiete des jordanischen Staates. *Zeitschrift des Deutschen Palästina-Vereins* 72:109–22.

Barker, G. 2000. Farmers, herders and miners in the Wadi Faynan, southern Jordan: A 10,000-year landscape archaeology. In *The archaeology of arid lands: Living at the margin*, ed. G. Barker and D. Gilbertson, 63–85. London: Routledge.

Barker, G. W., R. Adams, O. H. Creighton, D. Crook, D. D. Gilbertson, J. P. Grattan, C. O. Hunt, D. J. Mattingly, S. J. McLaren, H. A. Mohammed, P. Newson, C. Palmer, F. B. Pyatt, T. E. G. Reynolds, R. Tomber. 1999. Environment and land use in the Wadi Faynan, southern Jordan: The third season of geoarchaeology and landscape archaeology (1998). *Levant* 31:255–92.

Barker, G. W., R. Adams, O. H. Creighton, D. D. Gilbertson, J. P. Grattan, C. O. Hunt, D. J. Mattingly, S. J. McLaren, H. A. Mohammed, P. Newson, T. E. G. Reynolds, and D. C. Thomas. 1998. Environment and land use in the Wadi Faynan, southern Jordan: The second season of geoarchaeology and landscape archaeology (1997). *Levant* 30:5–25.

Barker, G. W., O. H. Creighton, D. D. Gilbertson, C. O. Hunt, D. J. Mattingly, S. J. McLaren, and D. C. Thomas. 1997. The Wadi Faynan Project, southern Jordan: A preliminary report on geomorphology and landscape archaeology. *Levant* 29:19–40.

Bar-Matthews, M., A. Ayalon, and A. Kaufman. 1998. Middle to late Holocene (6500 years period) paleoclimate in the eastern Mediterranean from isotopic composition of speleothems from Soreq Cave, Israel. In *Water, environment and society in times of climate change*, ed. A. Issar and N. Brown, 203–14. Dodrecht: Kluwer Academic.

———. 2000. Timing and hydrological conditions of sapropel events in the eastern Mediterranean, as evident from speleothems, Soreq cave, Israel. *Chemical Geology* 169 (1–2): 145–56.

Bar-Matthews, M., A. Ayalon, A. Kaufman, and G. J. Wasserburg. 1999. The eastern Mediterranean paleoclimate as a reflection of regional events: Soreq Cave, Israel. *Earth and Planetary Science Letters* 166 (1–2): 85–95.

Bartov, Y., M. Stein, Y. Enzel, A. Agnon, and Z. Reches. 2002. Lake levels and sequence stratigraphy of Lake Lisan, the Late Pleistocene precursor of the Dead Sea. *Quaternary Research* 57 (1): 9–21.

Baruch, U. 1990. Palynological evidence of human impact on the vegetation as recorded in Late Holocene lake sediments in Israel. In *Man's role in the shap-*

ing of the eastern Mediterranean landscape, ed. S. Bottema, G. Entjes-Nieborg, and W. van Zeist, 283–93. Rotterdam: Balkema.

———. 1993. *The palynology of late Holocene cores from the Dead Sea.* Ph.D. dissertation, Hebrew University of Jerusalem. (In Hebrew with an English summary)

———. 1999. The contribution of palynology and anthracology to archaeological research in the Levant. In *The practical impact of science on Near Eastern and Aegean archaeology*, ed. S. Pike and S. Gitin, 17–28. London: Archetype Publications for the Wiener Laboratory.

Baruch, U., and S. Bottema. 1991. Palynological evidence for climatic changes in the Levant ca. 17,000–9,000 B.P. In *The Natufian culture in the Levant*, ed. O. Bar-Yosef and F. R. Valla, 11–20. Ann Arbor, Mich.: International Monographs in Prehistory.

———. 1999. A new pollen diagram from Lake Hula: Vegetational, climatic, and anthropogenic implications. In *Ancient lakes: Their cultural and biological diversity*, ed. H. Kawanabe, G. W. Coulter, and A. C. Roosevelt, 75–86. Ghent, Belgium: Kenobi Productions.

Bar-Yosef, O. 1994. The Lower Paleolithic of the Near East. *Journal of World Prehistory* 8 (3): 211–63.

———. 1998. The Natufian culture in the Levant, threshold to the origins of agriculture. *Evolutionary Anthropology* 6 (5): 159–77.

———. 2000. The middle and early Upper Paleolithic in Southwest Asia and neighboring regions. In *The geography of Neandertals and modern humans in Europe and the Greater Mediterranean*, ed. O. Bar-Yosef and D. Pilbeam, 107–56. Cambridge, Mass.: Peabody Museum of Archaeology and Ethnology, Harvard University.

Bar-Yosef, O., and A. Belfer-Cohen. 1992. From foraging to farming in the Mediterranean Levant. In *Transition to agriculture in prehistory*, ed. A. B. Gebauer and T. D. Price, 21–48. Madison, Wisc.: Prehistory Press.

Bar-Yosef, O., and R. Kra, eds. 1994. *Chronology and paleoclimates of the eastern Mediterranean.* Tucson: *Radiocarbon*, University of Arizona.

Bar-Yosef, O., and R. H. Meadow. 1995. The origins of agriculture in the Near East. In *Last hunters–first farmers: New perspectives on the prehistoric transition to agriculture*, ed. T. D. Price and A. B. Gebauer, 39–94. Santa Fe, N.Mex.: School of American Research Press.

Bar-Yosef, O., and F. R. Valla. 1991. *The Natufian culture in the Levant.* Ann Arbor, Mich.: International Monographs in Prehistory.

Baubron, J. C., J. Besançon, J., L. Copeland, F. Hours, J. J. Macaire, and P. Sanlaville. 1985. Évolution de la moyenne vallée du Zarqa (Jordanie) au Néogène et au Quaternaire. *Revue de Géologie Dynamique et de Géographie Physique* 26 (5): 273–83.

Beaumont, P., and K. Atkinson. 1969. Soil erosion and conservation in northern Jordan. *Journal of Soil and Water Conservation* 24 (4): 144–47.

Begin, Z. B., W. Broecker, B. Buchbiner, Y. Druckman, A. Kaufman, M. Magaritz, and D. Neev. 1985. *Dead Sea and Lake Lisan levels in the last 30,000 years.* Report 85/29. Jerusalem: Geological Survey of Israel.

Beheiry, S. A. 1968–69. Geomorphology of central East Jordan. *Bulletin de la Société de Géographie d'Égypte* 41/42:5–22.

Beinart, W., and P. Coates. 1995. *Environment and history: The taming of nature in the USA and South Africa.* London: Routledge.

Bender, F. 1968. *Geologie von Jordanien.* Beitrage sur regionaled Geologie der Erde 7. Berlin: Gebrüder Borntraeger.

———. 1974. *Geology of Jordan.* Berlin: Gebrüder Borntraeger.

Besançon, J., B. Geyer, and P. Sanlaville. 1989. Contribution to the study of the geomorphology of the Azraq Basin, Jordan. In *The hammer on the rock: Studies in the Early Palaeolithic of Azraq, Jordan,* ed. L. Copeland and F. Hours. BAR International Series 540 (part 1), 7–64. Oxford: British Archaeological Reports.

Besançon, J., and F. Hours. 1985. Prehistory and geomorphology in northern Jordan: A preliminary outline. In *Studies in the history and archaeology of Jordan 2,* ed. A. Hadidi, 59–66. Amman: Department of Antiquities of Jordan.

Besançon, J., and P. Sanlaville. 1988. L'évolution geomorphologique du basin d'Azraq (Jordanie) depuis le pléistocene moyen. *Paléorient* 14 (2): 23–30.

Betts, A. V. G. 1991. The Jawa area in prehistory. In *Excavations at Jawa, 1972–1986: Stratigraphy, pottery and other finds,* ed. A. V. G. Betts, 181–90. Edinburgh, U.K.: Edinburgh University Press.

———. 1998. Holocene cultural ecology and environments of the northeastern Badia. In *The prehistoric archaeology of Jordan,* ed. D. O. Henry, 149–61. BAR International Series 705. Oxford: British Archaeological Reports.

Birkeland, P. W. 1999. *Soils and geomorphology.* 3rd ed. New York: Oxford University Press.

Blaikie, P., and H. Brookfield. 1987. Defining and debating the problem. In *Land degradation and society,* ed. P. Blaikie and H. Brookfield, 1–26. New York: Methuen.

Blanchet, G., P. Sanlaville, and M. Traboulsi. 1997. Le Moyen-Orient de 20,000 and BP à 6,000 ans BP essai de reconstitution paléoclimatique. *Paléorient* 32 (2): 187–86.

Bottema, S. 1975. The interpretation of pollen spectra from prehistoric settlements (with special attention to Liguliflorae). *Palaeohistoria* 57:17–35.

———. 1991. Pollen proxy data from southeastern Europe and the Near East. In *Evaluation of climate proxy data in relation to the European Holocene,* ed. B. Frenzel, A. Pons, and B. Glässer, 63–79. Stuttgart: Gustav Fischer.

————. 1995. The Younger Dryas in the eastern Mediterranean. *Quaternary Science Reviews* 14 (9): 883–91.

————. 1997. Third millennium climate change in the Near East based upon pollen evidence. In *Third millennium BC climate change and Old World collapse*, ed. H. N. Dalfes, G. Kukla, and H. Weiss, 489–515. Berlin: Springer-Verlag.

Bottema, S., and Y. Barkoudah. 1979. Modern pollen precipitation in Syria and Lebanon and its relation to vegetation. *Pollen et Spores* 21 (4): 427–80.

Bottema, S., G. Entjes-Nieborg, and W. van Zeist, eds. 1990. *Man's role in the shaping of the eastern Mediterranean landscape*. Rotterdam: Balkema.

Bottema, S., and H. Woldring. 1990. Anthropogenic indicators in the pollen record of the eastern Mediterranean. In *Man's role in the shaping of the eastern Mediterranean landscape*, ed. S. Bottema, G. Entjes-Nieborg, and W. van Zeist, 231–64. Rotterdam: Balkema.

Boulos, L. 1977. Studies on the flora of Jordan 5: On the flora of El-Jafr-Bayir Desert. *Candollea* 32:99–110.

Bourke, S. J. 2001. The Chalcolithic period. In *The archaeology of Jordan*, ed. B. MacDonald, R. Adams, and P. Bienkowski, 107–62. Sheffield, U.K.: Sheffield Academic Press.

Bradley, R. S. 1999. *Paleoclimatology: Reconstructing climates of the Quaternary.* 2nd ed. San Diego, Calif.: Academic Press.

Bruins, H. J. 1994. Comparative chronology of climate and human history in the southern Levant from the Late Chalcolithic to the Early Arab period. In *Late Quaternary chronology and paleoclimates of the eastern Mediterranean*, ed. O. Bar-Yosef and R. S. Kra, 301–14. Tucson: *Radiocarbon*, University of Arizona; Cambridge, Mass.: American School of Prehistoric Research.

————. 2001. Near East chronology: Towards an integrated ^{14}C time foundation. *Radiocarbon* 43 (3): 1147–54.

Bruins, H. J., I. Carmi, and E. Boaretto, eds. 2001. *Near East chronology: Archaeology and environment*. Special issue in *Radiocarbon* 43(3).

Bruins, H. J., and D. H. Yaalon. 1979. Stratigraphy of the Netivot section in the desert loess of the Negev (Israel). *Acta Geologica Academiae Scientiarum Hungaricae* 22:161–79.

Buehrle, A. 1992. *Analysis of phytoliths and paleoclimatic change at Wadi Ziqlab, Jordan.* M.A. thesis, University of Toronto.

Bull, W. B. 1991. *Geomorphic responses to climatic change.* New York: Oxford University Press.

Burckhardt, J. L. 1822. *Travels in Syria and the Holy Land.* Edited by William Martin Leake for the Association for Promoting the Discovery of the Interior Parts of Africa. London: John Murray.

Burdon, D. J. 1959. *Handbook of the geology of Jordan.* Amman: Government of the Hashemite Kingdom of Jordan.

Butzer, K. W. 1982. *Archaeology as human ecology.* Cambridge, U.K.: Cambridge University Press.

———. 1989. Cultural ecology. In *Geography in America,* ed. G. L. Gaile and C. J. Willmott, 192–208. Columbus, Ohio: Merrill Publishing Company.

———. 1992. The Americas before and after 1492: An introduction to current geographical research. *Annals of the Association of American Geographers* 82 (3): 345–68.

———. 1995. Environmental change in the Near East and human impact on the land. In *Civilizations of the Ancient Near East,* vol. 1, ed. Jack M. Sasson, 123–51. New York: Charles Scribner's Sons.

———. 1997. Sociopolitical discontinuity in the Near East c. 2200 B.C.E.: Scenarios from Palestine and Egypt. In *Third millennium BC climate change and Old World collapse,* ed. H. N. Dalfes, G. Kukla, and H. Weiss, 245–96. Berlin: Springer-Verlag.

Byrd, B. F. 1998. Spanning the gap from the Upper Paleolithic to the Natufian: The Early and Middle Paleolithic. In *The prehistoric archaeology of Jordan,* ed. D. O. Henry, 64–82. BAR International Series 705. Oxford: British Archaeological Reports.

Byrd, B. F., and S. M. Colledge. 1991. Early Natufian occupation along the edge of the southern Jordanian steppe. In *The Natufian culture in the Levant,* ed. O. Bar-Yosef and F. R. Valla, 265–76. Ann Arbor, Mich.: International Monographs in Prehistory.

Capaccioni, B., R. Franchi, O. Vaselli, E. Moretti, and F. Tassi. 2003. The origin of thermal waters from the eastern flank of the Dead Sea Rift Valley (western Jordan). *Terra Nova* 15:145–54.

Chapman, J. D. 1947. The forests of Transjordan. *Empire Forestry Review* 26 (2): 245–52.

Clark, G. A., N. R. Coinman, and M. P. Neeley. 2001. The Paleolithic of Jordan in the Levantine context. *Studies in the history and archaeology of Jordan 7,* ed. Department of Antiquities of Jordan staff, 49–68. Amman: Department of Antiquities of Jordan.

Clark, G. A., J. Lindley, M. Donaldson, A. Garrard, N. Coinman, J. Schuldenrein, S. Fish, and D. Olszewski. 1988. Excavations at Middle, Upper, and Epipaleolithic sites in the Wadi Hasa, west-central Jordan. In *The prehistory of Jordan: The state of research in 1986,* ed. A. N. Garrard and H. G. Gebel, 209–85. BAR International Series 396. Oxford: British Archaeological Reports.

Clark, G. A., J. Schuldenrein, M. L. Donaldson, H. P. Schwarcz, W. J. Rink, and S. K. Fish. 1997. Chronostratigraphic contexts of Middle Paleolithic horizons at the ʿAin Difla Rockshelter (WHS 634), west-central Jordan. In *The prehistory of Jordan 2: Perspectives from 1997,* ed. H. G. K. Gebel, Z. Kafafi, and G. O. Rollefson, 7–10. Berlin: Ex Oriente.

Clutton-Brock, J. 1981. *Domesticated animals from early times*. Austin: University of Texas Press.

Cohen, R., and W. G. Dever, 1981. Preliminary report of the third and final season of the Central Negev Highlands Project. *Bulletin of the American Schools of Oriental Research*, 243, 55–77.

Coinman, N. R. 1998. The Upper Paleolithic of Jordan. In *The prehistoric archaeology of Jordan*, ed. D. O. Henry, 39–63. BAR International Series 705. Oxford: British Archaeological Reports.

Colman, S. M., K. L. Pierce, and P. W. Birkeland. 1987. Suggested terminology for Quaternary dating methods. *Quaternary Research* 28:314–19.

Copeland, L. 1997. Status and future goals in Jordanian Paleolithic research. In Gebel, Kafafi, and Rollefson 1997, 183–191. Berlin: Ex Oriente.

Copeland, L., and F. Hours. 1988. The Paleolithic in north central Jordan: An overview of survey results from the Upper Zarqa and Azraq 1982–1986. In Garrard and Gebel 1988, 287–309. BAR International Series 396 (part 2). Oxford: British Archaeological Reports.

Copeland, L., and C. Vita-Finzi. 1978. Archaeological dating of geological deposits in Jordan. *Levant* 10:10–25.

Cordova, C. E. 1999a. Geoarchaeology of alluvial deposits and soils in the wadi systems of Northern Moab. *American Journal of Archaeology* 103 (3): 488–89.

———. 1999b. Landscape transformation in the Mediterranean-steppe transition zone of Jordan: A geoarchaeological approach. *Arab World Geographer* 2 (3): 188–201.

———. 2000. Geomorphological evidence of intense prehistoric soil erosion in the highlands of Central Jordan. *Physical Geography* 21 (6): 538–67.

———. 2005a. The degradation of the ancient Near Eastern environment. In *A companion to the ancient Near East*, ed. D. C. Snell, 109–25. Malden, Mass.: Blackwell.

———. 2005b. *Abstracts of the 101st Annual Meeting of the American Association of Geographers, Denver, Colorado*, 102. Washington, D.C.: Association of American Geographers.

Cordova, C. E., C. Foley, A. Nowell, and M. Bisson. 2005. Landforms, sediments, soil development and prehistoric site settings in the Madaba-Dhiban Plateau, Jordan. *Geoarchaeology* 20 (1): 29–56.

Cordova, C. E., and P. H. Lehman. 2003. Archaeopalynology of synanthropic vegetation in the chora of Chersonesos, Crimea, Ukraine. *Journal of Archaeological Science* 30 (11): 1483–1501.

Crawford, P. 1986. Flora of Tell Hesban and area, Jordan. In *Environmental foundations: Hesban 3*, ed. Ø. S. LaBianca and L. Lacelle, 75–98. Berrien Springs, Mich.: Andrews University Press.

Cullen, H. M., P. B. deMenocal, S. Hemming, G. Hemming, F. H. Brown, T. Guilderson, and F. Sirocko. 2000. Climate change and the collapse of the Akkadian empire: Evidence from the deep sea. *Geology* 28 (4): 379–82.

Dalfes, H. N., G. Kukla, and H. Weiss, eds. 1997. *Third millennium BC climate change and Old World collapse*. Berlin: Springer-Verlag.

Dallman, P. R. 1998. *Plant life in the world's Mediterranean climates: California, Chile, South Africa, Australia and the Mediterranean basin*. Berkeley: University of California Press.

Dan, J. 1977. The distribution and origin of nari and other lime crusts in Israel. *Israel Journal of Earth Sciences* 26:68–83.

Danin, A. 1983. *Desert vegetation of Israel and Sinai*. Jerusalem: Cana Publishing.

Davies, C. P. 2000. *Reconstruction of paleoenvironments from lacustrine deposits of the Jordan Plateau*. Ph.D. dissertation, Arizona State University, Tempe.

———. 2005. Quaternary paleoenvironments and potential for human exploitation of the Jordan Plateau desert interior. *Geoarchaeology* 20 (4): 379–400.

Davies, C. P., and P. L. Fall. 2001. Modern pollen precipitation from an elevational transect in central Jordan and its relationship to vegetation. *Journal of Biogeography* 28 (10): 1195–1210.

De Miroschedji, P., ed. 1989. *L'Urbanisation de la Palestine à l'Âge du Bronze Ancien*. BAR International Series 527. Oxford: British Archaeological Reports.

Denevan, W. M. 1992. The pristine myth: The landscape of the Americas in 1492. In *The Americas before and after 1492: Current geographical research*, ed. K. W. Butzer. *Annals of the Association of American Geographers* 82 (3): 369–85.

Dollfus, G., Z. Kafafi, J. Rewerski, N. Vaillant, E. Coquegniot, J. Desse, and R. Neef. 1988. Abu Hamid, an early fourth millennium site in the Jordan Valley. In Garrard and Gebel 1988, 567–77. BAR International Series 396. Oxford: British Archaeological Reports.

Donahue, J. 1981. Geologic investigations at Early Bronze sites. In *The Southeast Dead Sea Plain Expedition: An interim report of the 1977 season*, ed. W. E. Rast and R. T. Schaub, 137–54. Cambridge, Mass.: American Schools of Oriental Research.

———. 1984. Geologic reconstruction of Numeira. *Bulletin of the American Schools of Oriental Research* 255:83–88.

———. 1985. Hydrologic and topographic change during and after Early Bronze occupation at Bab edh-Dhra and Numeira. In *Studies in the history and archaeology of Jordan 2*, ed. A. Hadidi, 131–40. Amman: Department of Antiquities of Jordan.

———. 1988. Geologic history of Wadi el-Hasa survey area. In *The Wadi el-Hasa Archaeological Survey, 1979–1983, west-central Jordan*, ed. B. MacDonald, 26–39. Waterloo, Ont.: Wilfrid Laurier University Press.

Donahue, J., B. Peer, and R. T. Schaub. 1997. The Southeastern Dead Sea Plain: Changing shorelines and their impact on settlement patterns through historical periods. In *Studies in the history and archaeology of Jordan 6*, ed. G. Bisheh,

M. Zaghloul, and I. Kehrberg, 127–36. Amman: Department of Antiquities of Jordan.

Doughty, C. M. 1936. *Travels in Arabia Deserta*. New York: Random House.

Edwards, P. C., S. E. Falconer, P. L. Fall, I. Berelov, J. Czarzasty, C. Day, J. Meadows, C. Meegan, G. Sayej, T. K. Swoveland, and M. Westaway. 2004. Archaeology of the preliminary results of the second season of investigations by the joint La Trobe University/Arizona State University Project. *Annual of the Department of Antiquities of Jordan* 46:51–91.

Edwards, P. C., J. Meadows, G. Sayej, and M. Metzger. 2002. Zahrat adh-Dhraʿ2: A new Pre-Pottery Neolithic A site on the Dead Sea Plain in Jordan. *Bulletin of the American Schools of Oriental Research* 327:1–15.

El-Moslimanny, A. P. 1990. Ecological significance of common non-arboreal pollen: Examples from the drylands of the Middle East. *Review of Palaeobotany and Palynology* 64 (1–4): 343–50.

El-Radaideh, N. M. 1993. *Origin, composition and texture of the travertine in Wadi Haufa and Deir Abu Said areas*. Master's thesis, Yarmouk University, Irbid.

Emery-Barbier, A. 1995. Pollen analysis: Environmental and climatic implications. In *Prehistoric cultural ecology and evolution: Insights from southern Jordan*, ed. D. O. Henry, 375–84. New York: Plenum.

Emiliani, C. 1972. Quaternary paleotemperatures and the duration of the high temperature intervals. *Science* 178 (4059): 398–401.

Engel, T. 1993. Charcoal remains from an Iron Age copper smelting slag heap at Feinan, Wadi Arabah (Jordan). *Vegetation History and Archaeobotany* 2:205–11.

Enzel, Y., G. Kadan, and Y. Eyal. 2000. Holocene earthquakes inferred from a fan-delta sequence in the Dead Sea graben. *Quaternary Research* 53 (1): 34–48.

Evenari, M., L. Shanan, and N. Tadmor. 1982. *The Negev: The challenge of a desert*. Cambridge, Mass.: Harvard University Press.

Falconer, S., and P. Fall. 1995. Human impacts on the environment during the rise and collapse of civilization in the eastern Mediterranean. In *Late Quaternary environments and deep prehistory: A tribute to Paul S. Martin*, ed. D. W. Steadmand and J. I. Mead, 84–101. Hot Springs: Mammoth Site of Hot Springs, South Dakota.

Fall, P. 1990. Deforestation in southern Jordan: Evidence from fossil hyrax middens. In *Man's role in the shaping of the eastern Mediterranean landscape*, ed. S. Bottema, G. Entjes-Nieborg, and W. van Zeist, 271–81. Rotterdam: Balkema.

Fall, P., C. A. Lindquist, and S. E. Falconer. 1990. Fossil hyrax middens from the Middle East: A record of paleovegetation and human disturbance. In *Packrat middens: The last 40,000 years of biotic change*, ed. J. L. Betancourt, T. R. Van Devender, and P. S. Martin, 408–27. Tucson: University of Arizona Press.

Fall, P., L. Lines, and S. E. Falconer. 1998. Seeds of civilization: Bronze Age rural economy and ecology in the southern Levant. *Annals of the Association of American Geographers* 88 (1): 107–25.

Farhan, Y., S. Beheiry, and M. Abu-Safat. 1989. *Geomorphological studies on southern Jordan*. Amman: Publications of the University of Jordan. (In Arabic)

Faulkner, H., and A. Hill. 1997. Forests, soils and the threat of desertification. In *The Mediterranean environment and society*, ed. R. King, L. Proudfoot, and B. Smith, 252–72. London: Arnold.

Feinbrun, N., and M. Zohary. 1955. A geobotanical survey of Transjordan. *Bulletin of the Research Council of Israel* 5D:5–35.

Field, J. 1989. Geological setting at Beidha (Appendix A). In *The Natufian encampment at Beidha: Late Pleistocene adaptation in the southern Levant*, ed. B. F. Byrd, 86–96. Århus: Jutland Archaeological Society.

Field, J., and E. B. Banning. 1998. Hillslope processes and archaeology in Wadi Ziqlab, Jordan. *Geoarchaeology* 13 (6): 596–616.

Finnegan, M. 1981. Faunal remains from Bab edh-Dhra and Numeira. *Annual of the American Schools of Oriental Research* 46:177–80.

Fish, S. K. 1989. The Beidha pollen record (Appendix B). In *The Natufian encampment at Beidha*, ed. B. F. Byrd, 91–96. Århus: Jutland Archaeological Society.

Fontugne, M., M. Arnold, L. Labeyrie, M. Paterne, S. Calvert, and J. C. Duplessy. 1994. Paleoenvironment, sapropel chronology and Nile River discharge during the last 20,000 years as indicated by deep-sea sediment records in the eastern Mediterranean. In *Late Quaternary chronology and paleoclimates of the eastern Mediterranean*, ed. O. Bar-Yosef and R. Kra, 75–88. Tucson: *Radiocarbon*, University of Arizona.

Food and Agriculture Organization of the United Nations. 1988. *Soil map of the world: Revised legend*. World Soil Resources Report 60. Rome: Food and Agriculture Organization of the United Nations.

Frey, W., and H. Kürschner. 1989. *Vorderer Orient: Vegetation 1:8 000 000* (map A VI 1). Wiesbaden: Tübinger Atlas des Vorderen Orients.

Frumkin, A., I. Carmi, I. Zak, and M. Magaritz. 1994. Middle Holocene environmental change determined from the salt caves of Mount Sedom, Israel. In *Late Quaternary chronology of the eastern Mediterranean*, ed. Bar-Yosef and R. Kra, 315–22. Tucson: University of Arizona Press.

Frumkin, A., and Y. Elitzur. 2002. Historic Dead Sea level fluctuations calibrated with geological and archaeological evidence. *Quaternary Research* 57 (3): 334–42.

Garfunkel, Z. 1997. The history and formation of the Dead Sea basin. In *The Dead Sea: The lake and its setting*, ed. R. M. Niemi, Z. Ben Avraham, and J. R. Gat, 36–56. New York: Oxford University Press.

Garrard, A. N., D. Baird, and B. Byrd. 1994. The chronological basis and significance of the Late Paleolithic and Neolithic sequence in the Azraq Basin, Jordan. In *Late Quaternary chronology and paleoclimates of the eastern Mediterranean*, ed. O. Bar-Yosef and R. Kra, 177–99. Tucson: *Radiocarbon*, University of Arizona.

Garrard, A. N., A. Betts, B. Byrd, S. Colledge, and C. Hunt. 1988. A summary of paleoenvironmental and prehistoric investigation in the Azraq Basin. In *The prehistory of Jordan: The state of research in 1986*, ed. A. N. Garrard and H. G. Gebel, 311–37. BAR International Series 396. Oxford: British Archaeological Reports.

Garrard, A., S. Colledge, and L. Martin. 1996. The emergence of crop cultivation and caprine herding in the "Marginal Zone" of the southern Levant. In *The origins and spread of agriculture and pastoralism in Eurasia*, ed. D. R. Harris, 204–26. London: UCL Press.

Garrard, A. N., and H. G. Gebel, eds. 1988. *The prehistory of Jordan: The state of research in 1986*. BAR International Series 396, 2 vols. Oxford: British Archaeological Reports.

Gebel, H. G. K., H. D. Bienert, T. Krämer, B. Müller-Neuhof, R. Neef, J. Timm, and K. I. Wright. 1997. Ba'ja hidden in the Petra Mountains: Preliminary report on the 1997 excavations. In *The prehistory of Jordan 2: Perspectives from 1997*, ed. H. G. K. Gebel, Z. Kafafi, and G. O. Rollefson, 221–62. Berlin: Ex Oriente.

Gebel, H. K., Z. Kafafi, and G. O. Rollefson, eds. 1997. *The prehistory of Jordan 2: Perspectives from 1997*. Berlin: Ex Oriente.

Gilbert, A. S. 1995. The flora and fauna of the ancient Near East. In *Civilizations of the ancient Near East*, vol. 1, ed. J. M. Sasson, 153–74. New York: Charles Scribner's Sons.

Gilliland, D. R. 1986. Paleoethnobotany and paleobotany. In *Environmental foundations: Hesban 3*, ed. Ø. S. LaBianca and L. Lacelle, 121–42. Berrien Springs, Mich.: Andrews University Press.

Glueck, N. 1939. *Explorations in eastern Palestine 3*. Annual of American Schools of Oriental Research 18/19. New Haven, Conn.: American Schools of Oriental Research.

———. 1970. *The other side of the Jordan*. Cambridge, Mass.: American Schools of Oriental Research.

Goldberg, P. 1986. Late Quaternary environmental history of the southern Levant. *Geoarchaeology* 1 (3): 225–44.

———. 1994. Interpreting Late Quaternary continental sequences in Israel. In *Late Quaternary chronology and paleoclimates of the eastern Mediterranean*, ed. O. Bar-Yosef and R. Kra, 89–102. Tucson: *Radiocarbon*, University of Arizona.

Goldberg, P., and O. Bar-Yosef. 1990. The effect of man on geomorphological process based upon evidence from the Levant and adjacent areas. In *Man's role in the shaping of the eastern Mediterranean landscape*, ed. S. Bottema, G. Entjes-Nieborg, and W. van Zeist, 71–86. Rotterdam: Balkema.

Goldberg, P., V. T. Holliday, and C. R. Ferring, eds. 2001. *Earth sciences and archaeology*. New York: Kluwer Academic/Plenum.

Goodfriend, G. A. 1999. Terrestrial stable isotope records of Late Quaternary paleoclimates in the eastern Mediterranean region. *Quaternary Science Reviews* 18 (4–5): 501–13.

Goodfriend, G. A., and M. Magaritz. 1988. Paleosols and late Pleistocene fluctuations in the Negev Desert. *Nature* 332:144–46.

Gophna, R., N. Liphschitz, and S. Lev-Yadun. 1986. Man's impact on the natural vegetation of the central coastal plain of Israel during the Chalcolithic period and the Bronze Age. *Tel Aviv* 13:69–82.

Goring-Morris, N. A., and P. Goldberg. 1990. Late Quaternary dune incursions in the southern Levant: Archaeology, chronology and paleoenvironments. *Quaternary International* 5:113–37.

Goudie, A. S., P. Migon, R. J. Allison, and N. Rosser. 2002. Sandstone geomorphology of the Al-Quwayra area of south Jordan. *Zeitschrift für Geomorphologie* 46 (3): 365–90.

Grigg, D. B. 1974. *Agricultural systems of the world*. Cambridge, U.K.: Cambridge University Press.

Grove, A. T. 1997. Classics in physical geography revisited. *Progress in Physical Geography* 21 (2): 251–56.

Harris, D. R., ed. 1996. *The origins and spread of agriculture and pastoralism in Eurasia*. London: UCL Press.

Harrison, T. P. 1997. Shifting patterns of settlement in the highlands of central Jordan during the Early Bronze Age. *Bulletin of the American Schools of Oriental Research* 306:1–37.

Harrison, T. P., B. Hesse, S. H. Savage, and D. W. Schnurrenberger. 2000. Urban life in the highlands of central Jordan: A preliminary report of the 1996 Tall Madaba excavations. *Annual of the Department of Antiquities of Jordan* 44:211–30.

Hassan, F. A. 1995. Late Quaternary geology and geomorphology of the area in the vicinity of Ras en Naqb. In *Prehistoric cultural ecology and evolution: Insights from southern Jordan*, ed. D. O. Henry, 23–31. New York: Plenum.

Hatough-Bouran, A. M., D. M. H. Al-Eisawi, and A. M. Disi. 1986. The effect of conservation on wildlife in Jordan. *Environmental Conservation* 13 (4): 331–35.

Hatough-Bouran, A. M., and A. M. Disi. 1991. History, distribution, and conservation of large mammals and their habitats in Jordan. *Environmental Conservation* 18:28–44.

Hatough-Bouran, A. M., M. A. Duwayri, A. M. Disi, Z. S. Amr, R. E. Nasr, A. M. Budieri, and I. A. Al-Khader. 1998. *Jordan country study on biological diversity*. Amman: General Corporation for the Environment Protection.

Hauptmann, A., and G. Weisgerber. 1987. Archaeometallurgical and mining archaeological investigations in the area of Feinan, Wadi ʿArabah (Jordan). *Annual of the Department of Antiquities of Jordan* 31:419–37.

Head, M. J. 1999. Radiocarbon dating of arid zone deposits. In *Paleoenvironmental reconstruction in arid lands*, ed. A. K. Singvhi and E. Derbyshire, 293–326.
Rotterdam: A. Balkema.

Heim, C., N. R. Nowaczyk, J. F. W. Negendank, S. A. G. Leroy, and Z. Ben-
Avraham. 1997. Near East desertification: Evidence from the Dead Sea. *Natur-
wissenschaften* 84 (9): 398–401.

Helms, S. W. 1981. *Jawa: Lost city of the Black Desert*. Ithaca, N.Y.: Cornell University Press.

Hembleben, C., D. Mesichner, R. Zahn, A. Almogi-Labin, H. Erlenkeuser, and
B. Hiller. 1996. Three hundred eighty thousand year long stable isotope and
faunal records from the Red Sea: Influence of global sea level change on hy-
drography. *Paleoceanography* 11 (2): 147–56.

Henry, D. O. 1989. *From foraging to agriculture: The Levant at the end of the Ice
Age*. Philadelphia: University of Pennsylvania.

———. 1995. *Prehistoric cultural ecology and evolution: Insights from southern
Jordan*. New York: Plenum.

———. 1997a. Cultural and geologic successions of Middle and Upper Paleolithic deposits in the Jebel Qalkha area of southern Jordan. In *The prehistory
of Jordan 2: Perspectives from 1997*, ed. H. G. K. Gebel, Z. Kafafi, and G. O.
Rollefson, 69–76. Berlin: Ex Oriente.

———. 1997b. Prehistoric human ecology in the southern Levant east of the rift
from 20,000–6,000 BP. *Paléorient* 23 (2): 107–19.

———, ed. 1998a. *The prehistoric archaeology of Jordan*. BAR International
Series 705. Oxford: British Archaeological Reports.

———. 1998b. Introduction. In *The prehistoric archaeology of Jordan*, ed. D. O.
Henry, 1–4. BAR International Series 705. Oxford: British Archaeological Reports.

———, ed. 2003a. *Neanderthals in the Levant: Behavioral organization and the
beginnings of human modernity*. London: Continuum.

———. 2003b. A case study from southern Jordan: Tor Faraj. In *Neanderthals in
the Levant: Behavioral organization and the beginnings of human modernity*,
ed. D. O. Henry, 33–59. London: Continuum.

Henry, D. O., H. A. Bauer, K. W. Kerry, J. E. Beaver, and J. J. White. 2001. Survey
of prehistoric sites, Wadi Araba, southern Jordan. *Bulletin of the American
Schools of Oriental Research* 323:1–19.

Henry, D. O., C. Cordova, J. J. White, R. M. Dean, J. E. Beaver, H. Ekstrom,
S. Kadowaki, J. McCorriston, A. Nowell, and L. Scott-Cummings. 2003. The
Early Neolithic site of Ayn Abū Nukhayla, southern Jordan. *Bulletin of the
American Schools of Oriental Research* 330:1–30.

Henry, D. O., S. Hall, H. Hietala, Y. Demidenko, V. Usik, A. Rosen, and P. Thomas. 1996. Middle Paleolithic behavioral organization: 1993 excavation of Tor Faraj, southern Jordan. *Journal of Field Archaeology* 23 (1): 31–53.

Hill, J. B. 2002. *Land use and land abandonment: A case study from the Wadi al-Hasa, west-central Jordan.* Ph.D. dissertation, Arizona State University, Tempe.

Hillman, G. C. 1996. Late Pleistocene changes in wild plant-foods available to hunter-gatherers of the northern Fertile Crescent: Possible preludes to cereal cultivation. In *The origins and spread of agriculture and pastoralism in Eurasia*, ed. D. R. Harris, 159–203. London: UCL Press.

Holliday, V. T., ed. 1992. *Soils in archaeology: Landscape evolution and human occupation.* Washington, D.C.: Smithsonian Institution Press.

Horowitz, A. 1969. Recent pollen sedimentation in Lake Kinneret, Israel. *Pollen et spores* 11 (2): 353–84.

———. 1979. *The Quaternary of Israel.* New York: Academic Press.

———. 1992. *Palynology of arid lands.* Amsterdam: Elsevier.

Horowitz, A., M. Weinstein, and E. Ganor. 1975. Palynological determination of dust storm's provenances in Israel. *Pollen et spores* 17 (2): 223–31.

Huckriede, R., and G. Wiesemann. 1968. Der jungpleistozäne Pluvial-See von El Jafr und weitere Daten zum Quartär Jordaniens. *Geologica et Palaeontologica* 2:73–95.

Hulings, N. C., and M. I. Wahbeh. 1988. *A guide to the sea shore of Jordan.* Amman: University of Jordan Press.

Hunt, C. O., H. A. Elrishi, D. D. Gilbertson, J. Grattan, S. McLaren, F. B. Pyatt, G. Rushworth, and G. W. Barker. 2004. Early-Holocene environments in Wadi Faynan, Jordan. *Holocene* 14 (6): 921–30.

Huntington, E. 1911. *Palestine and its transformation.* Boston: Houghton Mifflin; New York: Cambridge University Press.

Huntley, B., and H. J. B. Birks. 1983. *An atlas of past and present pollen maps for Europe: 0–13,000 years ago.* London: Cambridge University Press.

Ishida, S., A. Parker, D. Kennet, and M. J. Hodson. 2003. Phytolith analysis from the archaeological site of Kush, Ras al-Khaimah, United Arab Emirates. *Quaternary Research* 59 (3): 310–21.

Issar, A. S. 1990. *Water shall flow from the rock.* Heidelberg: Springer-Verlag.

———. 2003. *Climate changes during the Holocene and their impact on hydrological systems.* International Hydrological Series. Cambridge, U.K.: Cambridge University Press.

Jacobsen, T., and R. M. Adams. 1958. Salt and silt in ancient Mesopotamian agriculture. *Science* 128 (3334): 1251–58.

Johns, J., A. McQuitty, R. Falkner, and project staff. 1989. The Faris Project: Preliminary report upon the 1986 and 1988 seasons. *Levant* 21:63–95.

Jordan Meteorological Data (JMD). 2005. Climate Data. http://www.jmd.gov.jo (last accessed 1 December 2005).

Josephus, Flavius. 1968. *Josephus in nine volumes*, vol. 3: *The Jewish wars, books IV–VII*, trans. H. S. J. Thackeray. Cambridge, Mass.: Harvard University Press.

Kafafi, Z. 1992. Pottery Neolithic settlement patterns in Jordan. In *Studies in the history and archaeology of Jordan 4*, ed. S. Tell, 115–23. Amman: Department of Antiquities of Jordan.

——. 1998. The Late Neolithic in Jordan. In *The prehistoric archaeology of Jordan*, ed. D. O. Henry, 127–38. BAR International Series 705. Oxford: British Archaeological Reports.

Kafafi, Z., G. Palumbo, A. A. Al-Shiyab, F. Parenti, E. Santucci, M. Hatmaleh, M. Shunnaq, and M. Wilson. 1997. The Wadi Az-Zarqaʿ/Wadi ad-Dulayl Archaeological Project: Report on the 1996 fieldwork season. *Annual of the Department of Antiquities of Jordan* 41:9–26.

Kahana, R., B. Ziv, Y. Enzel, and U. Dayan. 2002. Synoptic climatology of major floods in the Negev Desert, Israel. *International Journal of Climatology* 22 (7): 867–82.

Kaufman, A., G. J. Wasserburg, D. Porcelli, M. Bar-Matthews, A. Ayalon, and L. Halicz. 1998. U-Th isotope systematics from the Soreq Cave, Israel, and climatic correlations. *Earth Planetary Science Letters* 156 (3–4): 141–55.

Kaufman, A., Y. Yechieli, and M. Gardosh. 1992. Reevaluation of the lake-sediment chronology of the Dead Sea basin, Israel, based on ^{230}Th/U dates. *Quaternary Research* 38 (3): 292–304.

Kedar, A. Z. 1985. The Arab conquests and agriculture: A seventh century apocalypse, satellite imagery, and palynology. *Asian and African Studies* 19:1–15.

Kelso, G. K., and G. O. Rollefson. 1989. Two late Quaternary pollen profiles from Ain El-Assad, Azraq, Jordan. In *The hammer on the rock: Studies in the Early Palaeolithic of Azraq, Jordan*, ed. L. Copeland and F. Hours, 259–75. BAR International Series 540 (part 1). Oxford: British Archaeological Reports.

Kennedy, D., and D. Riley. 1990. *Rome's desert frontier from the air*. Austin: University of Texas Press.

Khoury, F. 1998. Habitat associations and communities of breeding birds in the highlands of south-west Jordan. *Zoology in the Middle East* 16:35–48.

Khoury, H., E. Salameh, and P. Udluft. 1984. On the Zerka Maʿin (Therma Kallirrhoes) Travertine/Dead Sea (hydrochemistry, geochemistry and isotopic composition). *Neues Jahrbuch für Geologie und Paläontologie* 8:472–84.

Khresat, S. A. 2001. Calcic horizon distribution and soil classification in selected soils on north-western Jordan. *Journal of Arid Environments* 47 (2): 145–52.

Khresat, S. A., Z. Rawajfih, and M. Mohamad. 1998. Morphological, physical and chemical properties of selected soils in the arid and semiarid region in north-western Jordan. *Journal of Arid Environments* 40 (1): 15–25.

King, R. 1997. Introduction: An essay on Mediterraneanism. In *The Mediterranean environment and society*, ed. R. King, L. Proudfoot, and B. Smith, 1–11. London: Arnold.

Kislev, M. E. 1990. Extinction of *Acacia nilotica* in Israel: A methodological approach. In *Man's role in the shaping of the eastern Mediterranean landscape*, ed. S. Bottema, G. Entjes-Nieborg, and W. Van Zeist, 307–17. Rotterdam: A. Balkema.

Köhler-Rollefson, I. 1988. The aftermath of the Levantine Neolithic revolution in the light of ecological and ethnographic evidence. *Paléorient* 14 (1): 87–93.

———. 1997. Proto-élevage, pathologies, and pastoralism: A post-mortem of the process of goat domestication. In *The prehistory of Jordan 2: Perspectives from 1997*, ed. H. G. K. Gebel, Z. Kafafi, and G. O. Rollefson, 557–65. Berlin: Ex Oriente.

Köhler-Rollefson, I., L. Quintero, and G. Rollefson. 1993. A brief note on the fauna from Neolithic 'Ain Ghazal. *Paléorient* 19 (2): 95–98.

Köhler-Rollefson, I., and G. Rollefson. 1990. The impact of Neolithic subsistence strategies on the environment: The case of 'Ain Ghazal, Jordan. In *Man's role in the shaping of the eastern Mediterranean landscape*, ed. S. Bottema, G. Entjes-Nieborg, and W. Van Zeist, 3–14. Rotterdam: A. Balkema.

Kronfeld, J., J. Vogel, E. Rosenthal, and M. Weinstein-Evron. 1988. Age and paleoclimatic implications of the Bet Shean travertines. *Quaternary Research* 30 (3): 298–303.

Krupp, F., and W. Schneider. 1989. The fishes of Jordan River drainage basin and Azraq oasis. *Fauna of Saudi Arabia* 10:347–416.

Küchler, A. W. 1988. The nature of vegetation. In *Vegetation mapping*, ed. A. W. Küchler and I. S. Zonneveld, 13–23. Dordrecht: Kluwer Academic Publishers.

Kuijt, I. 2000. People and space in early agricultural villages: Exploring daily lives, community size and architecture in the late Pre-Pottery Neolithic. *Journal of Anthropological Archaeology* 19 (1): 75–12.

Kürschner, H. 1986. A physiognomical-ecological classification of the vegetation of southern Jordan. In *Contributions to the vegetation of Southwest Asia: Beihefte zum Tübinger Atlas des Vorderen Orients*. Reihe A (Naturwissenchaften) 24, ed. H. Kürschner, 45–79. Wiesbaden: Ludwig Reichert.

Kutiel, P. 1994. Fire and ecosystem heterogeneity: A Mediterranean case study. *Earth Surface Processes and Landforms* 19 (2): 187–94.

Kutiel, P., and Z. Naveh. 1987. Soil properties beneath *Pinus halepensis* and *Quercus calliprinos* trees on burned and unburned mixed forest on Mt. Carmel, Israel. *Forest Ecology and Management* 20:11–24.

LaBianca, Ø. S. 1990. *Sedentarization and nomadization: Food system cycles at Hesban and vicinity in Transjordan*. Hesban 1. Berrien Springs, Mich.: Institute of Archaeology and Andrews University Press.

LaBianca, Ø. S., and A. Von den Driesch, eds. 1995. *Faunal remains: Taphonomi-*

cal and zooarchaeological studies of the animal remains from Tell Hesban and vicinity. Hesban Final Reports 13. Berrien Springs, Mich.: Andrews University Press.

Lacelle, L. 1986. Bedrock, surficial geology, and soils. In *Environmental foundations: Studies of climatical, ceological, hydrological, and phytological conditions in Hesban and vicinity,* ed. Ø. S. LaBianca and L. Lacelle, 23–58. Hesban 2. Berrien Springs, Mich.: Andrews University Press.

Landmann, G., G. M. Abu Qudaira, K. Shawabkeh, V. Wrede, and S. Kempe. 2002. Geochemistry of the Lisan and Damya formations in Jordan and implications for paleoclimate. *Quaternary International* 89 (1): 45–57.

Leroi-Ghouran, A. 1982. Palynological research of Near Eastern archaeological sites. In *Paleoclimates, palaeoenvironments and human communities in the eastern Mediterranean region in later prehistory,* ed. J. L. Bintliff and W. Van Zeist, 353–56. BAR International Series 133 (part 2). Oxford: British Archaeological Reports.

Leroi-Ghouran, A., and F. Darmon. 1991. Analyses polliniques de stations natoufiennes au Proche Orient. In *The Natufian culture in the Levant,* ed. O. Bar-Yosef and F. R. Valla, 21–26. Ann Arbor, Mich.: International Monographs in Prehistory.

Levy, T., R. Adams, and R. Shafiq. 1999. The Jebel Hamrat Fidan Project: Excavations at the Wadi Fidan 40 cemetery, Jordan (1997). *Levant* 31:293–308.

Lewis, N. N. 1987. *Nomads and settlers in Syria and Jordan, 1800–1980.* Cambridge, U.K.: Cambridge University Press.

Liphschitz, N. 1986. Overview of the dendrochronological and dendroarchaeological research in Israel. *Dendrochronologia* 4:37–58.

———. 1987. *Ceratonia siliqua* in Israel: Ancient element or a newcomer? *Israel Journal of Botany* 36 (4): 191–97.

Liphschitz, N., and G. Biger. 2001. Past distribution of Aleppo pine (*Pinus halepensis*) in the mountains of Israel (Palestine). *Holocene* 11 (4): 427–36.

Liphschitz, N., R. Gophna, M. Hartman, and G. Biger. 1991. The beginning of olive (*Olea europaea*) cultivation in the Old World: A reassessment. *Journal of Archaeological Science* 18 (4): 441–53.

Long, G. A. 1957. *The bioclimatology of eastern Jordan.* Technical Report. Rome: Food and Agriculture Organization of the United Nations.

Lowdermilk, W. C. 1944. *Palestine: Land of promise.* New York: Harper and Brothers.

Lowenthal, D. 2000. *George Perkins Marsh: Prophet of conservation.* Seattle: University of Washington Press.

Mabry, J. 1992. *Alluvial cycles and early agricultural settlement phases in the Jordan Valley.* Ph.D. dissertation, University of Arizona, Tucson.

MacDonald, B. 2000. *East of the Jordan: Territories and sites of the Hebrew scriptures.* Boston: American Schools of Oriental Research.

MacDonald, B., R. Adams, and P. Bienkowski, eds. 2001. *The archaeology of Jordan*. Sheffield, U.K.: Sheffield Academic Press.

Machlus, M., Y. Enzel, S. L. Goldstein, S. Marco, and M. Stein. 2000. Reconstructing low levels of Lake Lisan by correlating fan-delta and lacustrine deposits. *Quaternary International* 73/74:137–44.

Macumber, P. G. 2001. Evolving landscape and environment in Jordan. In *The archaeology of Jordan*, ed. B. MacDonald, R. Adams, and P. Bienkowski, 1–30. Sheffield, U.K.: Sheffield Academic Press.

Macumber, P. G., and P. C. Edwards. 1997. Preliminary results from the Acheulian site of Mashari'a, and a new stratigraphic framework for the Lower Palaeolithic of the East Jordan Valley. In *The prehistory of Jordan 2: Perspectives from 1997*, ed. H. G. K. Gebel, Z. Kafafi, and G. O. Rollefson, 23–43. Berlin: Ex Oriente.

Macumber, P. G., and M. J. Head. 1991. Implications of the Wadi al-Hammeh sequences for the terminal drying of Lake Lisan, Jordan. *Palaeogeography, Palaeoclimatology, Palaeoecology* 84 (1–4): 163–73.

Magaritz, M., A. Kaufman, and D. H. Yaalon. 1981. Calcium carbonate nodules in soils: $^{18}O/^{16}O$ and $^{13}C/^{12}C$ ratios and ^{14}C contents. *Geoderma* 25 (3–4): 157–72.

Maher, L., and E. B. Banning. 2001. Geoarchaeological survey in Wādi Ziqlāb, Jordan. *Annual of the Department of Antiquity of Jordan* 45:61–70.

———. 2002. Geoarchaeological survey and the Epipaleolithic in northern Jordan. *Antiquity* 76 (292): 313–14.

Mandel, R. D., and A. H. Simmons. 1988. A preliminary assessment of the geomorphology of 'Ain Ghazal. In *The prehistory of Jordan: The state of research in 1986*, ed. A. N. Garrard and H. G. Gebel, 431–36. BAR International Series 396. Oxford: British Archaeological Reports.

McCorriston, J., and F. Hole. 1991. The ecology of seasonal stress and the origin of agriculture in the Near East. *American Anthropologist* 93 (1): 46–69.

Meadows, J. 2005. The Younger Dryas episode and the radiocarbon chronologies of the Lake Huleh and Ghab Valley pollen diagrams, Israel and Syria. *Holocene* 15 (4): 631–36.

Meissner, R. 1986. *The continental crust: A geophysical approach*. Orlando, Fla.: Academic Press.

Merrill, S. 1883. *East of the Jordan: A record of travel and observation in the countries of Moab, Gilead and Bashan during the years 1875–1877*. New York: Charles Scribners Sons.

Metzger, M. C. 1984. Faunal remains at Tell-el-Hayyat: Preliminary results. *Bulletin of the American Schools of Oriental Research* 255:68–69.

Miller, N. 1997. The macrobotanical evidence for vegetation in the Near East c. 18,000/16,000 BC to 4000 BC. *Paléorient* 23 (2): 197–207.

Millington, A., S. al-Hussein, and R. Dutton. 1999. Population dynamics, socioeconomic change and land colonization in northern Jordan, with special reference to the Badia Research and Development Project area. *Applied Geography* 19 (4): 363–84.

Mitrakos, K. 1980. A theory for Mediterranean plant life. *Acta Oecologica* 1 (3): 245–52.

Moore, A. M. T., and G. C. Hillman. 1992. The Pleistocene to Holocene transition and human economy in southwest Asia: The impact of the Younger Dryas. *American Antiquity* 57 (3): 482–94.

Moorman, F. 1959. *The soils of East Jordan: Report to the government of Jordan.* Expanded Technical Assistance Program 1132. Rome: Food and Agriculture Organization (FAO) of the United Nations.

Moumani, K., J. Alexander, and M. D. Bateman. 2003. Sedimentology of the Late Quaternary Wadi Hasa Marl Formation of central Jordan: A record of climate variability. *Palaeogeography, Palaeoclimatology, Palaeoecology* 191 (2): 221–42.

Mountfort, G. 1965. *Portrait of a desert: The story of an expedition to Jordan.* Boston: Houghton Mifflin.

Mousa, S. 1982. Jordan: Towards the end of the Ottoman Empire, 1841–1918. In *Studies in the archaeology and history of Jordan 1,* ed. A. Hadidi, 385–91. Amman: Department of Antiquities of Jordan.

Muheisen, M. 1988. A survey of prehistoric sites in the Jordan Valley. In *The prehistory of Jordan: The state of research in 1986,* ed. A. N. Garrard and H. G. Gebel, 503–6. BAR International Series 396. Oxford: British Archaeological Reports.

Munro, R. N., R. V. R. Morgan, and W. J. Jobling. 1997. Optical dating and landscape chronology at ad-Disa, southern Jordan and its potential. In *Studies in the history and archaeology of Jordan 6,* ed. G. Bisheh, M. Zaghloul, and I. Kehrberg, 97–103. Amman: Department of Antiquities of Jordan.

Naveh, Z., and J. Dan. 1973. The human degradation of Mediterranean landscapes in Israel. In *Mediterranean type ecosystems: Origin and structure,* ed. F. Di Castri and H. A. Mooney, 373–90. New York: Springer-Verlag.

Neef, R. 1990. Introduction, development and environmental implications of olive culture: The evidence from Jordan. In *Man's role in the shaping of the eastern Mediterranean landscape,* ed. S. Bottema, G. Entjes-Nieborg, and W. van Zeist, 295–306. Rotterdam: Balkema.

———. 1997. Status and perspectives of archaeobotanical research in Jordan. In *The prehistory of Jordan 2: Perspectives from 1997,* ed. H. G. K. Gebel, Z. Kafafi, and G. O. Rollefson, 601–9. Berlin: Ex Oriente.

Neeley, M. P., J. D. Peterson, G. A. Clark, S. K. Fish, and M. Glass. 1998. Investigations at Tor al-Tareeq: An Epipaleolithic site in the Wadi el-Hasa, Jordan. *Journal of Field Archaeology* 25 (3): 295–317.

Neev, D., and K. O. Emery. 1995. *The destruction of Sodom, Gomorrah, and*

Jericho: Geological, climatological, and archaeological background. New York: Oxford University Press.

Neev, D., and J. K. Hall. 1979. Geophysical investigations in the Dead Sea. *Sedimentary Geology* 23 (1–4): 209–38.

Nelson, B. 1974. *Azraq: Desert oasis.* Athens: Ohio University Press.

Niemi, T. M. 1997. Fluctuations of Late Pleistocene Lake Lisan in the Dead Sea Rift. In *The Dead Sea: The lake and its setting,* ed. R. M. Niemi, Z. Ben Avraham, and J. R. Gat, 226–36. New York: Oxford University Press.

Niemi, T. M., Z. Ben-Avraham, and J. R. Gat. 1997. Dead Sea research — An introduction. In *The Dead Sea: The lake and its setting,* ed. R. M. Niemi, Z. Ben Avraham, and J. R. Gat, 3–7. New York: Oxford University Press.

Niemi, T. M., J. B. J. Harrison, H. Zhang, M. Attalan, and J. Bruce. 2001. Late Pleistocene and Holocene slip rate of the northern Wadi Araba fault, Dead Sea transform, Jordan. *Journal of Seismology* 5 (3): 449–74.

Niemi, T. M., and A. M. Smith II. 1999. Initial results of the southeastern Wadi Araba, Jordan Geoarchaeological Study: Implications for shifts in Late Quaternary aridity. *Geoarchaeology* 14 (8): 791–820.

Niklewski, J., and W. van Zeist. 1970. A Late Quaternary pollen diagram from northwestern Syria. *Acta Botanica Neerlandica* 19:737–54.

Obeidat, O. 1992. *Geochemistry, mineralogy, and petrography of the travertine of Deir Alla and Zerqaʿ Maʿin hot springs, West Jordan.* Master's thesis. Yarmouk University, Irbid, Jordan.

Oleson, J. P. 1997. Landscape and cityscape in the Hisma: The resources of ancient Al-Humayma. In *Studies in the history and archaeology of Jordan 6,* ed. M. Zaghloul and G. Bisheh, 175–88. Amman: Department of Antiquities.

Olszewski, D. I. 1997. From the Late Ahmarian to the Early Natufian: A summary of hunter-gatherer activities at Yutil Al-Hasa, west-central Jordan. In *The prehistory of Jordan 2: Perspectives from 1997,* ed. H. G. K. Gebel, Z. Kafafi, and G. O. Rollefson, 171–83. Berlin: Ex Oriente.

———. 2001. The Paleolithic period, including the Epipaleolithic. In *The archaeology of Jordan,* ed. B. MacDonald, R. Adams, and P. Bienkowski, 31–65. Sheffield, U.K.: Sheffield Academic Press.

Olszewski, D., and N. Coinman. 2002. An ice age oasis in Jordan: Pleistocene hunter-gatherers in the Wadi al-Hasa region. *Expedition* 44:16–23.

Olszewski, D., and J. B. Hill. 1997. Renewed excavations at Tabaqa (WHS 895), an Early Natufian site in the Wadi al-Hasa, Jordan. *Neolithics: A Newsletter of Southwest Asian Lithics Research* 3/97:11–12.

Osborn, G., and J. M. Duford. 1981. Geomorphological processes in the inselberg region of south-western Jordan. *Palestine Exploration Quarterly* 113:1–17.

Palmer, C. 1998. "Following the plough": The agricultural environment of northern Jordan. *Levant* 30:129–65.

Palumbo, G. 2001. The Early Bronze Age IV. In *The archaeology of Jordan,* ed.

B. MacDonald, R. Adams, and P. Bienkowski, 233–69. Sheffield, U.K.: Sheffield Academic Press.

Parenti, F., A. H. Al-Shiyab, E. Santucci, Z. Kafafi, G. Palumbo, and C. Guérin. 1997. Early Acheulean stone tools and fossil faunas from the Dauqara Formation, Upper Zarqa Valley, Jordanian Plateau. In *The prehistory of Jordan 2: Perspectives from 1997*, ed. H. G. K. Gebel, Z. Kafafi, and G. O. Rollefson, 7–22. Berlin: Ex Oriente.

Parker, A. G., L. Eckersley, M. M. Smith, A. S. Goudie, S. Stokes, S. Ward, K. White, and J. Hodson. 2004. Holocene vegetation dynamics in the northeastern Rubʿ al Khali desert, Arabian Peninsula: A phytolith, pollen and carbon isotope study. *Journal of Quaternary Science* 19 (7): 665–76.

Philip, G. 2001. The Early Bronze I–III ages. In *The archaeology of Jordan*, ed. B. MacDonald, R. Adams, and P. Bienkowski, 163–232. Sheffield, U.K.: Sheffield Academic Press.

Pick, W. P. 1990. Meissner Pasha and the construction of railways in Palestine and neighboring countries. In *Ottoman Palestine, 1800–1914: Studies in economic and social history*, ed. G. G. Gilbar, 177–218. Leiden: E. J. Brill.

Pignatti, S. 1978. Evolutionary trends in Mediterranean flora and vegetation. *Vegetatio* 37 (3): 175–85.

Pollard, M., ed. 1999. *Geoarchaeology: Exploration, environments and resources.* London: Geological Society.

Pons, A. 1981. The history of the Mediterranean shrublands. In *Mediterranean-type shrublands*, ed. F. Di Castri, D. W. Goodall, and R. L. Specht, 131–38. Ecosystems of the World 11. Amsterdam: Elsevier.

Prag, K. 1984. Continuity and migration in the South Levant in the late third millennium. *Palestine Exploration Quarterly* 116:58–86.

Pye, K., and D. Sherwin, 1999. Loess. In *Aeolian environments, sediments and landforms*, ed. A. S. Goudie and S. Stokes, 213–38. New York: John Wiley and Sons.

Quintero, L. A. 1996. Flint mining in the Pre-Pottery Neolithic: Preliminary report on the exploitation of flint at Neolithic ʿAin Ghazal in highland Jordan. In *Neolithic chipped stone industries of the Fertile Crescent and their contemporaries in adjacent regions*, ed. S. K. Kozlowski and H. G. K. Gebel, 233–42. Berlin: Ex Oriente.

Quintero, L. A., P. J. Wilke, and G. O. Rollefson. 2002. From flint mine to fan scraper: The late prehistoric Jafr Industrial Complex. *Bulletin of the American Schools of Oriental Research* 327:17–48.

Rapp, G., Jr., and C. L. Hill. 1998. *Geoarchaeology: The earth-science approach to archaeological interpretation.* New Haven, Conn.: Yale University Press.

Raven, P. H. 1971. The relationships between "Mediterranean" floras. In *Plant life of South-West Asia*, ed. P. H. Davis, P. C. Harper, and I. C. Hedge, 119–34. Edinburgh, U.K.: Botanical Society of Edinburgh.

Redford, D. O. 1992. *Egypt, Canaan, and Israel in ancient times.* Princeton, N.J.: Princeton University Press.

Reifenberg, A. 1955. *The struggle between the desert and the sown: Rise and fall of agriculture in the Levant.* Jerusalem: Government Press.

Richard, S. 1980. Toward a consensus of opinion on the end of the Early Bronze Age in Palestine-Transjordan. *Bulletin of the American Schools of Oriental Research* 237:5–34.

———. 2000. Chronology vs. regionalism in the Early Bronze IV period: An assemblage of whole and restored vessels from the public building at Khirbet Iskander. In *The archaeology of Jordan and beyond: Essays in honor of James A. Sauer,* ed. L. E. Stager, J. A. Greene, and M. D. Coogan, 399–417. Winona Lake, Ind.: Eisenbrauns.

Richard, S., and C. J. Long. 1995. Archaeological expedition to Khirbet Iskander and its vicinity. *Annual of the Department of Antiquities of Jordan* 39:81–92.

Richard, S., C. J. Long, and B. Libby. 2001. Khirbet Iskander. *American Journal of Archaeology* 105 (3): 440–41.

Rikli, M. A. 1943. *Das Pflanzenkleid des Mittelmeerländer,* 3 vols. Bern: H. Huber.

Roberts, N. 1982. Forest re-advance and the Anatolian Neolithic. In *Archaeological aspects of woodland ecology,* ed. M. Bell and S. Limbrey, 231–46. BAR International Series 146. Oxford: British Archaeological Reports.

Roberts, N., and H. E. Wright Jr. 1993. The Near East and Southwest Asia. In *Global climatic changes since the Last Glacial Maximum,* ed. H. E. Wright Jr., J. E. Kutzbach, T. Webb III, W. F. Ruddiman, F. A. Street-Perrott, and P. J. Bartlein, 194–220. Minneapolis: University of Minnesota Press.

Rollefson, G. O. 1998. The aceramic Neolithic. In *The prehistoric archaeology of Jordan,* ed. D. O. Henry, 102–26. BAR International Series 705. Oxford: British Archaeological Reports.

———. 2000. Return to ʿAin el-Assad (Lion Spring), 1996: Azraq Acheulian occupation in situ. In *The archaeology of Jordan and beyond: Essays in honor of James A. Sauer,* ed. L. E. Stager, J. A. Greene, and M. D. Coogan, 418–428. Winona Lake, Ind.: Eisenbrauns.

———. 2001a. Jordan in the seventh and sixth millennia BC. *Studies in the history and archaeology of Jordan 7,* ed. Department of Antiquities of Jordan staff, 95–100. Amman: Department of Antiquities of Jordan.

———. 2001b. The Neolithic period. In *The archaeology of Jordan,* ed. B. Mac-Donald, R. Adams, and P. Bienkowski, 67–105. Sheffield, U.K.: Sheffield Academic Press.

Rollefson, G. O., D. Schnurrenberger, L. A. Quintero, R. P. Watson, and R. Low. 1997. ʿAin Soda and ʿAin Qasiya: New Late Pleistocene and Early Holocene sites in the Azraq Shishan area, eastern Jordan. In *The prehistory of Jordan 2: Perspectives from 1997,* ed. H. G. K. Gebel, Z. Kafafi, and G. O. Rollefson, 45–58. Berlin: Ex Oriente.

Roquero, C. 1993. Some problems on genesis and classification of red soils in Spain. *Second meeting on red Mediterranean soils* [proceedings], 21–42. Adana, Turkey: Soil Science Society of Türkiye.

Rosen, A. M. 1986. Environmental change and settlement at Tel Lachish, Israel. *Bulletin of the American Schools of Oriental Research* 263:55–60.

———. 1995. Preliminary analysis of phytoliths from prehistoric sites in southern Jordan. In *Prehistoric cultural ecology and evolution: Insights from southern Jordan*, ed. D. O. Henry, 399–403. New York: Plenum.

———. 1997a. Environmental change and human adaptational failure at the end of the Early Bronze Age in the southern Levant. In *Third millennium BC climate change and Old World collapse*, ed. H. N. Dalfes, G. Kukla, and H. Weiss, 25–38. Berlin: Springer-Verlag.

———. 1997b. The geoarchaeology of Holocene environments and land use at Kazane Höyük, S. E. Turkey. *Geoarchaeology* 12 (4): 395–416.

———. 2003. Middle Paleolithic plant exploitation: The microbotanical evidence. In *Neanderthals in the Levant: Behavioral organization and the beginnings of human modernity*, ed. D. O. Henry, 156–71. London: Continuum.

Rosen, A. M., and S. Wiener, 1994. Identifying ancient irrigation: A new method using opaline phytoliths from emmer wheat. *Journal of Archaeological Science* 21 (1): 125–32.

Rosen, S. A. 2003. Early multi-resource nomadism: Excavations at the Camel site in the central Negev. *Antiquity* 77 (298): 750–61.

Rosen, S. A., and G. Avni. 1993. The edge of the empire: The archaeology of pastoral nomads in the southern Negev highlands in late antiquity. *Biblical Archaeologist* 56:189–98.

Rösner, U. 1989. Löss am Rande der Wüstensteppe? *Erdkunde* 43 (4): 233–42.

Rossignol-Strick, M. 1995. Sea-land correlation of pollen records in the eastern Mediterranean for the glacial-interglacial transition: Biostratigraphy versus radiometric time scale. *Quaternary Science Reviews* 14 (9): 893–915.

———. 1998. Paléoclimat de la Méditerranée orientale et de l'Asie du Sud-Ouest de 15000 à 6000 BP. *Paléorient* 23 (2): 175–86.

———. 1999. The Holocene Climate Optimum and pollen records of Sapropel 1 in the eastern Mediterranean, 9000–6000 BP. *Quaternary Science Reviews* 18 (4–5): 515–30.

Rowe, A. G. 1999. The exploitation of an arid landscape by a pastoral society: The contemporary Eastern Badia of Jordan. *Applied Geography* 19 (4): 345–61.

Rowton, M. B. 1967. The woodlands of ancient western Asia. *Journal of Near Eastern Studies* 26:261–77.

Royal Jordanian Geographic Centre. 1984. *National atlas of Jordan*, part 1: *Climate and agroclimatology*. Amman: Royal Jordanian Geographic Centre.

Royal Society for the Conservation of Nature (RSCN). 2005. Protected Areas.

Royal Society for the Conservation of Nature. Amman, Jordan. http://www .rscn.org.jo/rscn/protectedarea.htm (last accessed 1 December 2005).

Russell, K. W. 1995. Traditional Bedouin agriculture at Petra: Ethnoarchaeological insights into the evolution of food production. In *Studies in the history and archaeology of Jordan 5*, ed. K. ʿAmr, F. Zayadine, and M. Zaghloul, 693–705. Amman: Department of Antiquities of Jordan.

Salibi, K. 1998. *The modern history of Jordan.* London: I. B. Tauris.

Sancetta, C., J. Imbrie, and N. G. Kipp. 1973. The climatic record of the past 130,000 years in the North Atlantic deep-sea core V23–83: Correlation with the terrestrial record. *Quaternary Research* 3 (1): 110–16.

Sanlaville, P. 1996. Changements climatiques dans la région Levantine à la fin du Pléistocène Supérieur et au début de l'Holocène, leurs relations avec l'évolution des sociétés humaines. *Paleoriént* 22 (1): 7–30.

Saqqa, W., and M. Atallah. 2004. Characterization of the aeolian terrain facies in Wadi Araba Desert, southwestern Jordan. *Geomorphology* 62 (1–2): 63–87.

Savage, S. 2005. Welcome to Doc Savage Archaeology Web Page. http:// archaeology.asu.edu/Jordan/index.html (last date accessed 1 December 2005).

Scholz, F., and G. Schweizer. 1992. *Vorderer Orient: Nomadismus und andere Formen der Wanderviehwirtschaft.* Part A: *Geography*, map A X-11, scale 1:8 million. Tübinger Atlas des Vorderen Orients (TAVO). University of Tübingen.

Schuldenrein, J., and G. A. Clark. 1994. Landscape and prehistoric chronology of west-central Jordan. *Geoarchaeology* 9 (1): 31–55.

———. 2001. Prehistoric landscapes and settlement geography along the Wadi Hasa, west-central Jordan: Geoarchaeology, human palaeoecology and ethnographic modelling, part 1. *Environmental Archaeology* 6:23–38.

———. 2003. Prehistoric landscapes and settlement geography along the Wadi Hasa, west-central Jordan: Geoarchaeology, human palaeoecology and ethnographic modelling, part 2. *Environmental Archaeology* 8:1–16.

Schuldenrein, J., and P. Goldberg. 1981. Late Quaternary palaeoenvironments and prehistoric site distribution in the lower Jordan Valley: A preliminary report. *Paléorient* 7 (1): 57–72.

Schumacher, G., L. Oliphant, and G. Le Strange. 1886. *Across the Jordan: Being an exploration and survey of part of Hauran and Jaulan.* New York: Scribner and Welford.

Schwab, M. J., F. Neumann, T. Litt, J. F. W. Negendank, and M. Stein. 2004. Holocene palaeoecology of the Golan Heights (Near East): Investigation of lacustrine sediments from Birkat Ram crater lake. *Quaternary Science Reviews* 23 (16–17): 1723–31.

Schyle, D., and H. G. K. Gebel. Upper Palaeolithic Siq Umm al-Alda 1, near Wadi Musa, southern Jordan. In *The prehistory of Jordan 2: Perspectives from 1997*, ed. H. G. K. Gebel, Z. Kafafi, and G. O. Rollefson, 149–70. Berlin: Ex Oriente.

Scott-Cummings, L. 2001. *Pollen, phytolith, and spherulite analysis of samples from Jordan.* Paleo Research Labs Technical Report 00–12 (unpublished). Golden, Colo.: Paleo Research Laboratories.

Seetzen, U. J. 1854. *Ulrich Jasper Seetzen's Reisen durch Syrien, Palästina, Phoenicien, die Transjordan-Länder, Arabia Petraea und Unter-Aegypten,* ed. F. Kruse. Berlin: G. Reimer.

Sellars, J. R. 1998. The Natufian of Jordan. In *The prehistoric archaeology of Jordan,* ed. D. O. Henry, 83–101. BAR International Series 705. Oxford: British Archaeological Reports.

Shackleton, N. J., and N. D. Opdyke, 1973. Oxygen isotope stratigraphy of equatorial Pacific core V28–238: Oxygen isotope temperatures and ice volume on a 105 and 106 year scale. *Quaternary Research* 3 (1): 39–95.

Shawabekeh, Kh. 1998. *The geology of Ma'in area.* Map sheet no. 3153, 3. Bulletin 40. Amman: Geology Directorate, Geological Mapping Division.

Shmida, A., and J. A. Aronson. 1986. Sudanian elements in the flora of Israel. *Annals of the Missouri Botanical Garden* 73:1–28.

Simmons, A. H., I. Köhler-Rollefson, G. O. Rollefson, R. Mandel, and Z. Kafafi. 1988. 'Ain Ghazal: A major Neolithic settlement in central Jordan. *Science* 240 (4848): 35–39.

Simmons, A., G. Rollefson, Z. Kafafi, R. Mandel, M. an-Nahar, J. Cooper, I. Kohler-Rollefson, and K. Durand. 2001. Wadi Shu'eib, a large Neolithic community in central Jordan: Final report of test investigations. *Bulletin of the American Schools of Oriental Research* 321:1–39.

Soil Survey and Land Research Centre. 1993. *National soil map and land use project: The soils of Jordan.* Main report. Amman: Ministry of Agriculture.

Soil Survey Staff. 1996. *Keys to soil taxonomy.* 7th ed. Washington, D.C.: U.S. Department of Agriculture/Natural Resources Conservation Service.

Stager, L. E. 1985. The first fruits of civilization. In *Palestine in the Bronze and Iron Ages: Papers in honour of Olga Tufnell,* ed. J. N. Tubb, 172–87. London: Institute of Archaeology, University of London.

Stein, J. K., and W. R. Farrand, eds. 2001. *Sediments in archaeological context.* Salt Lake City: University of Utah Press.

Stein, M. 2001. The sedimentary and geochemical record of Neogene-Quaternary water bodies in the Dead Sea basin — inferences. *Journal of Paleolimnology* 26 (3): 271–82.

Stein, M., A. Starinsky, A. Agnon, A. Katz, M. Raab, B. Spiro, and I. Zak. 2000. The impact of brine rock-interaction during marine evaporite formation on the isotopic Sr record in the oceans: Evidence from Mt. Sedom, Israel. *Geochimica et Cosmochimica Acta* 64 (12): 2039–53.

Steinitz, G., and Y. Bartov. 1992. The Miocene-Pleistocene history of the Dead Sea segment of the rift in light of K-Ar ages of basalts. *Israel Journal of Earth Sciences* 40 (1–4): 199–208.

Steuernagel, D. C. 1925. Der Adschlun. *Zeitschrift des Deutschen Palästina-Vereins* 48:1–201.

Stiller, M., A. Ehrlich, U. Baruch, and A. Kaufman. 1983–84. *The Late Holocene sediments of Lake Kinneret (Israel): Multidisciplinary study of a 5-m core.* Jerusalem: Geological Survey of Israel, Ministry of Energy and Infrastructure.

Suc, J. P. 1984. Origin and evolution of the Mediterranean vegetation and climate in Europe. *Nature* 307:429–32.

Taylor, J. 2002. *Petra and the lost kingdom of the Nabataeans.* Cambridge, Mass.: Harvard University Press.

Tchernov, E. 1987. The age of the 'Ubeidiya Formation, an Early Pleistocene hominid site in the Jordan Valley, Israel. *Israel Journal of Earth Sciences* 36 (1–2): 3–30.

———. 1988. The biogeographical history of the southern Levant. In *The zoogeography of Israel*, ed. Y. Yom-Tov and E. Tchernov, 159–250. Dodrecht: W. Junk Publishers.

———. 1991. Biological evidence for human sedentism in southwest Asia during the Natufian. In *The Natufian culture in the Levant*, ed. O. Bar-Yosef and F. R. Valla, 315–40. Ann Arbor, Mich.: International Monographs in Prehistory.

Tellawi, A. M. M. 2001. *Conservation and sustainable use of biological diversity in Jordan.* Amman: General Corporation for the Environment Protection.

Tomaselli, R. 1981. Relations with other ecosystems: Temperate evergreen forests, Mediterranean coniferous forests, savannahs, steppes and desert shrublands. In *Mediterranean-type shrublands*, ed. F. Di Castri, D. W. Goodall, and R. L. Specht, 123–30. Ecosystems of the World 11. Amsterdam: Elsevier.

Toplyn, M. R. 1987. Sampled faunal remains from the el-Lejjun barracks. In *The Roman frontier in central Jordan*, ed. S. T. Parker, 705–22. BAR International Series 340. Oxford: British Archaeological Reports.

Touchan, R., and M. K. Hughes. 1999. Dendrochronology in Jordan. *Journal of Arid Environments* 42 (4): 291–303.

Tristram, H. B. 1873. *The Land of Moab: Travels and discoveries on the east side of the Dead Sea and the Jordan.* New York: Harper.

Tsoar, H., V. Goldsmith, S. Schoenhoaus, and K. Clarke. 1995. Reversed desertification on sand dunes along the Sinai/Negev border. In *Desert aeolian processes*, ed. V. P. Tchakerian, 251–67. London: Chapman and Hall.

Vaks, A., M. Gilmour, C. J. Hawkesworth, A. Frumkin, A. Kaufman, A. Matthews, M. Bar-Matthews, A. Ayalon, and B. Schilman. 2003. Paleoclimate reconstruction based on the timing of speleothem growth and oxygen and carbon isotope composition in a cave located in the rain shadow of Israel. *Quaternary Research* 59 (2): 182–93.

Van Andel, T. H. 1998a. Middle and Upper Paleolithic environments and the calibration of 14C dates beyond 10,000 B.P. *Antiquity* 72:26–33.

————. 1998b. Paleosols, red sediment, and the Old Stone Age in Greece. *Geo-archaeology* 13 (4): 361–90.

Van Andel, T. H., and C. Runnels. 1987. *Beyond the Acropolis: A rural Greek past.* Stanford, Calif.: Stanford University Press.

Van der Plicht, J. 1999. Radiocarbon calibration for the Middle/Upper Palaeolithic: A comment. *Antiquity* 73 (279): 119–23.

Van der Plicht, J., and H. J. Bruins. 2001. Radiocarbon dating in Near Eastern contexts: Confusion and quality control. *Radiocarbon* 43 (3): 1155–66.

Van Zeist, W., and S. Bottema. 1991. *Late Quaternary vegetation of the Near East.* TAVO Beihefte Reihe A 18. Wiesbaden: Ludwig Reichert.

Van Zeist, W., H. Woldring, and D. Stapert. 1975. Late Quaternary vegetation and climate of southwestern Turkey. *Palaeohistoria* 17:53–143.

Vita-Finzi, C. 1964. Observations in the Late Quaternary of Jordan. *Palestine Exploration Quarterly* 96:19–33.

————. 1966. The Hasa Formation: An alluvial deposition in Jordan. *Man* 1:386–90.

————. 1969. *The Mediterranean valleys: Geological changes in historical times.* London: Cambridge University Press.

Vita-Finzi, C., and G. W. Dimbleby. 1971. Medieval pollen from Jordan. *Pollen et spores* 13 (3): 415–420.

Von den Driesch, A., and U. Wodtke. 1997. The fauna of ʿAin Ghazal, a major PPN and Early PN settlement in central Jordan. In *The prehistory of Jordan 2: Perspectives from 1997*, ed. H. G. K. Gebel, Z. Kafafi, and G. O. Rollefson, 511–56. Berlin: Ex Oriente.

Wagstaff, J. M. 1985. *The evolution of Middle Eastern landscapes: An outline to A.D. 1840.* London: Croom Helm.

Walker, B. J. 1999. Militarization to nomadization: The Middle and Late Islamic periods. *Near Eastern Archaeology* 62 (4): 202–33.

Waters, M. R. 1992. *Principles of geoarchaeology: A North American perspective.* Tucson: University of Arizona Press.

Weatherbase. 2005. Weather Records and Averages. Canty and Associates LLC. http://www.weatherbase.com (last accessed 1 December 2005).

Weinstein, M. 1976. The Late Quaternary vegetation of the northern Golan. *Pollen et spores* 18 (4): 553–62.

————. 1979. Airborne pollen. In *The Quaternary of Israel*, ed. A. Horowitz, 180–85. New York: Academic Press.

Weinstein-Evron, M. 1987. Palynology of Pleistocene from the Arava Valley, Israel. *Quaternary Research* 27 (1): 82–88.

Weiss, H. 2000. Beyond the Younger Dryas: Collapse as adaptation to abrupt climate change in ancient West Asia and the eastern Mediterranean. In *Confronting natural disaster: Engaging the past to understand the future*, ed.

G. Bawden and R. Reycraft, 75–98. Albuquerque: University of New Mexico Press.

Wigley, T. M. L., and G. Farmer. 1982. Climate of the eastern Mediterranean and Near East. In *Paleoclimates, palaeoenvironments and human communities in the eastern Mediterranean region in later prehistory*, ed. J. L. Bintliff and W. Van Zeist, 3–37. BAR International Series 133 (part 1). Oxford: British Archaeological Reports.

Wilkinson, T. J. 1994. The structure and dynamics of dry-farming states in Upper Mesopotamia. *Current Anthropologist* 35 (5): 483–520.

———. 1997. Environmental fluctuations, agricultural production and collapse: A view from Bronze Age Upper Mesopotamia. In *Third millennium BC climate change and Old World collapse*, ed. H. N. Dalfes, G. Kukla, and H. Weiss, 67–106. Berlin: Springer-Verlag.

———. 1998. The archaeological landscape of the upper Balikh Valley, Syria: Investigations from 1992–1995. *Journal of Field Archaeology* 25 (1): 63–87.

———. 1999. Holocene Valley fills of southern Turkey and northwestern Syria: Recent geoarchaeological contributions. *Quaternary Science Reviews* 18 (4–5): 555–71.

———. 2003. *Archaeological landscapes of the Near East*. Tucson: University of Arizona Press.

Willcox, G. 1999. Charcoal analysis and Holocene vegetation history in southern Syria. *Quaternary Science Reviews* 18 (4–5): 711–16.

Williams, M., D. Dunkerley, P. De Decker, P. Kershaw, and J. Chappell. 1998. *Quaternary environments*. 2nd ed. London: Arnold.

Wilson, C., C. Warren, C. R. Conder, H. H. K. Kitchener, E. H. Palmer, G. Smith, G. Chester, and M. Clermont-Ganneau. 1881. *The survey of western Palestine: Special papers on topography, manners and customs, etc.* London: Committee of the Palestine Exploration Fund.

Wright, H., Jr. 1993. Environmental determinism in eastern prehistory. *Current Anthropology* 34 (4): 458–69.

Yaalon, D. H. 1997. Soils in the Mediterranean region: What makes them different? *Catena* 28 (3–4): 157–69.

Yaalon, D. H., and E. Ganor. 1979. East Mediterranean trajectories of dust carrying storms from the Sahara and Sinai. In *Saharan dust: Mobilization, transport, deposition*, ed. C. Morales, 187–93. SCOPE Report 14. New York: John Wiley and Sons.

Yasuda, Y., H. Kitagawa, and T. Nakagawa. 2000. The earliest record of major anthropogenic deforestation in the Ghab Valley, northwest Syria: A palynological study. *Quaternary International* 73/74:127–36.

Yechieli, Y., M. Magaritz, Y. Levy, U. Weber, U. Kafri, W. Woelfli, and G. Bonani. 1993. Late Quaternary geological history of the Dead Sea area, Israel. *Quaternary Research* 39 (1): 59–67.

Zak, I. 1997. Evolution of the Dead Sea brines. In *The Dead Sea: The lake and its setting*, ed. R. M. Niemi, Z. Ben Avraham, and J. R. Gat, 133–44. New York: Oxford University Press.

Zohary, D. 1969. The progenitors of wheat and barley in relation to domestication and agriculture dispersal in the Old World. In *The domestication and exploitation of plants and animals*, ed. P. J. Ucko and G. W. Dimbleby, 47–66. London: Duckworth.

Zohary, D., and P. Spiegel-Roy. 1975. Beginning of fruit growing in the Old World. *Science* 187 (4174): 319–27.

Zohary, M. 1952. A monographical study of the genus *Pistacia*. *Palestine Journal of Botany* [Jerusalem Series] 5:187–228.

———. 1960. The maquis of *Quercus calliprinos* in Israel and Jordan. *Bulletin of the Research Council of Israel* 9D: 51–72.

———. 1973. *Geobotanical foundations of the Middle East*, 2 vols. Stuttgart: Gustav Fischer.

Zohary, M., and N. Feinbrun-Dothan. 1966. *Flora Palaestina*, 4 parts. Jerusalem: Israel Academy of Sciences and Humanities.

Zonneveld, I. S. 1988. The ITC method of mapping natural and seminatural vegetation. In *Vegetation mapping*, ed. A. W. Küchler and I. S. Zonneveld, 401–26. Dordrecht: Kluwer Academic Publishers.

Zuhair, A. n.d. Animal Biodiversity of Jordan. http://amon.nic.gov.jo/biodiversity (last accessed 1 December 2005).

Index

About the Author

Carlos E. Cordova is an associate professor of geography at Oklahoma State University in Stillwater. He obtained his bachelor's and master's degrees at the National Autonomous University of Mexico (UNAM) and his Ph.D. at the University of Texas at Austin.

His areas of expertise are geoarchaeology, geomorphology, and Quaternary palynology and phytoliths. His research focuses on environmental change during the Late Quaternary, focusing mainly on vegetation, humans, and climate. Although his main study areas are in Jordan and the Crimean Peninsula, he has participated in numerous research projects in the Near East, the eastern Mediterranean, the Black Sea region, the Great Plains of North America, southern Alaska, the American Southwest, and Mexico. His research has been published in *Geoarchaeology, Journal of Archaeological Science, Physical Geography, The Arab World Geographer, Quaternary International*, and *The Holocene*, among others.

Current research projects include vegetation reconstruction of the Mediterranean–Black Sea–Caspian Sea corridor during the past 30,000 years, as part of a cooperative project between the International Geological Correlation Project 521 and the Bristol Initiative for the Dynamic Global Environment (BRIDGE), geoarchaeology of Neandertal habitations in the Ma'in area of Jordan, reconstruction of ancient Greek colonization in the Black Sea, and eolian and alluvial paleoenvironments of the Dust Bowl–stricken area in the southern Great Plains of North America.